Biotechnological Innovations in Animal Productivity

BOOKS IN THE BIOTOL SERIES

The Molecular Fabric of Cells
Infrastructure and Activities of Cells

Techniques used in Bioproduct Analysis
Analysis of Amino Acids, Proteins and Nucleic Acids
Analysis of Carbohydrates and Lipids

Principles of Cell Energetics
Energy Sources for Cells
Biosynthesis and the Integration of Cell Metabolism

Genome Management in Prokaryotes
Genome Management in Eukaryotes

Crop Physiology
Crop Productivity

Functional Physiology
Cellular Interactions and Immunobiology
Defence Mechanisms

Bioprocess Technology: Modelling and Transport Phenomena
Operational Modes of Bioreactors

In vitro Cultivation of Micro-organisms
In vitro Cultivation of Plant Cells
In vitro Cultivation of Animal Cells

Bioreactor Design and Product Yield
Product Recovery in Bioprocess Technology

Techniques for Engineering Genes
Strategies for Engineering Organisms

Technological Applications of Biocatalysts
Technological Applications of Immunochemicals

Biotechnological Innovations in Health Care

Biotechnological Innovations in Crop Improvement
Biotechnological Innovations in Animal Productivity

Biotechnological Innovations in Energy and Environmental Management

Biotechnological Innovations in Chemical Synthesis

Biotechnological Innovations in Food Processing

Biotechnology Source Book: Safety, Good Practice and Regulatory Affairs

BIOTECHNOLOGY BY OPEN LEARNING

Biotechnogical Innovations in Animal Productivity

PUBLISHED ON BEHALF OF:

Open universiteit and **Thames Polytechnic**

Valkenburgerweg 167
6401 DL Heerlen
Nederland

Avery Hill Road
Eltham, London SE9 2HB
United Kingdom

Butterworth-Heinemann

Butterworth–Heinemann Ltd
Linacre House, Jordan Hill, Oxford OX2 8DP

PART OF REED INTERNATIONAL BOOKS

OXFORD LONDON BOSTON
MUNICH NEW DELHI SINGAPORE SYDNEY
TOKYO TORONTO WELLINGTON

First published 1992

British Library Cataloguing in Publication Data

A catalogue record for this book is available from the British Library

Library of Congress Cataloguing in Publication Data

A catalogue record for this book is available from the Library of Congress, Washington

ISBN 0 7506 1511 7

Composition by Thames Polytechnic
Printed and Bound in Great Britain by Thomson Litho Ltd, East Kilbride, Scotland

The Biotol Project

The BIOTOL team

OPEN UNIVERSITEIT, THE NETHERLANDS
Dr M. C. E. van Dam-Mieras
Professor W. H. de Jeu
Professor J. de Vries

THAMES POLYTECHNIC, UK
Professor B. R. Currell
Dr J. W. James
Dr C. K. Leach
Mr R. A. Patmore

This series of books has been developed through a collaboration between the Open universiteit of the Netherlands and Thames Polytechnic to provide a whole library of advanced level flexible learning materials including books, computer and video programmes. The series will be of particular value to those working in the chemical, pharmaceutical, health care, food and drinks, agriculture, and environmental, manufacturing and service industries. These industries will be increasingly faced with training problems as the use of biologically based techniques replaces or enhances chemical ones or indeed allows the development of products previously impossible.

The BIOTOL books may be studied privately, but specifically they provide a cost-effective major resource for in-house company training and are the basis for a wider range of courses (open, distance or traditional) from universities which, with practical and tutorial support, lead to recognised qualifications. There is a developing network of institutions throughout Europe to offer tutorial and practical support and courses based on BIOTOL both for those newly entering the field of biotechnology and for graduates looking for more advanced training. BIOTOL is for any one wishing to know about and use the principles and techniques of modern biotechnology whether they are technicians needing further education, new graduates wishing to extend their knowledge, mature staff faced with changing work or a new career, managers unfamiliar with the new technology or those returning to work after a career break.

Our learning texts, written in an informal and friendly style, embody the best characteristics of both open and distance learning to provide a flexible resource for individuals, training organisations, polytechnics and universities, and professional bodies. The content of each book has been carefully worked out between teachers and industry to lead students through a programme of work so that they may achieve clearly stated learning objectives. There are activities and exercises throughout the books, and self assessment questions that allow students to check their own progress and receive any necessary remedial help.

The books, within the series, are modular allowing students to select their own entry point depending on their knowledge and previous experience. These texts therefore remove the necessity for students to attend institution based lectures at specific times and places, bringing a new freedom to study their chosen subject at the time they need and a pace and place to suit them. This same freedom is highly beneficial to industry since staff can receive training without spending significant periods away from the workplace attending lectures and courses, and without altering work patterns.

Contributors

AUTHORS

Professor A. Brand, University of Utrecht, The Netherlands

Dr G.J. Garssen, Research Institute for Animal Production, Zeist, The Netherlands

Dr V.E. Hick, Manchester Polytechnic, Manchester, UK

Dr Th. A.M. Kruip, Embrytec bv., Zeist, The Netherlands

Dr P.J. van der Meer, Ministry of VROM, Leidschendam, The Netherlands

Professor J.T. van Oirshot, Central Veterinary Institute, Lelystad, University of Utrecht, The Netherlands

Dr J.K. Oldenbroek, Research Institute for Animal Production, Zeist, The Netherlands

Professor W.J.M. Spaan, University of Leiden, The Netherlands

EDITOR

Dr N.A.A. Macfarlane, Nottingham Polytechnic, Nottingham, UK

SCIENTIFIC AND COURSE ADVISORS

Dr M. C. E. van Dam-Mieras, Open universiteit, Heerlen, The Netherlands

Dr C. K. Leach, Leicester Polytechnic, Leicester, UK

ACKNOWLEDGEMENTS

Grateful thanks are extended, not only to the authors, editor and course advisors, but to all those who have contributed to the development and production of this book. They include Mrs A. Allwright, Dr M. de Kok, Miss J. Skelton and Professor R. Spier. Our thanks also to the many authors and publishers who allowed us to use published material to support this text. Especial thanks to the British Veterinary Association, Royal Netherlands Veterinary Association, American Veterinary Medical Association, Academic Press Inc. and Cold Spring Harbor Laboratory Press. The development of this BIOTOL text has been funded by COMETT, The European Community Action programme for Education and Training for Technology, by the Open universiteit of The Netherlands and by Thames Polytechnic.

Project Manager Dr J.W. James

Contents

How to use an open learning text

An open learning text presents to you a very carefully thought out programme of study to achieve stated learning objectives, just as a lecturer does. Rather than just listening to a lecture once, and trying to make notes at the same time, you can with a BIOTOL text study it at your own pace, go back over bits you are unsure about and study wherever you choose. Of great importance are the self assessment questions (SAQs) which challenge your understanding and progress and the responses which provide some help if you have had difficulty. These SAQs are carefully thought out to check that you are indeed achieving the set objectives and therefore are a very important part of your study. Every so often in the text you will find the symbol Π, our open door to learning, which indicates an activity for you to do. You will probably find that this participation is a great help to learning so it is important not to skip it.

Whilst you can, as a open learner, study where and when you want, do try to find a place where you can work without disturbance. Most students aim to study a certain number of hours each day or each weekend. If you decide to study for several hours at once, take short breaks of five to ten minutes regularly as it helps to maintain a higher level of overall concentration.

Before you begin a detailed reading of the text, familiarise yourself with the general layout of the material. Have a look at the contents of the various chapters and flip through the pages to get a general impression of the way the subject is dealt with. Forget the old taboo of not writing in books. There is room for your comments, notes and answers; use it and make the book your own personal study record for future revision and reference.

At intervals you will find a summary and list of objectives. The summary will emphasise the important points covered by the material that you have read and the objectives will give you a check list of the things you should then be able to achieve. There are notes in the left hand margin, to help orientate you and emphasise new and important messages.

BIOTOL will be used by universities, polytechnics and colleges as well as industrial training organisations and professional bodies. The texts will form a basis for flexible courses of all types leading to certificates, diplomas and degrees often through credit accumulation and transfer arrangements. In future there will be additional resources available including videos and computer based training programmes.

Preface

It is extremely difficult to predict the precise impact that biotechnology will have on agriculture. Despite this uncertainty, many believe that biotechnology represents a means for pivotal change leading to an extension of the scope and efficiency of agricultural production. This text and its companion BIOTOL volume 'Biotechnological Innovations in Crop Improvement' explain the application of biotechnology to improving agricultural productivity.

In this text, we examine the application of biotechnology to animal production. Animal production, of course, depends upon the reproductive capabilities of animals, their growth rates and the ability of the farmer and veterinary services to prevent and cure infection. This text focuses onto these facets of animal production by explaining how the growth and reproduction of livestock may be manipulated and how diagnostic systems and vaccines may be developed using contemporary biotechnology procedures. The whole tenor of the text is on the application of biotechnology and it has been assumed that the reader is familiar with the key stages in recombinant DNA technology and has a background knowledge of animal physiology, reproduction and metabolism. The sections on vaccines and diagnostics are written on the assumption that the reader has knowledge of the basic features of the immune response including the roles of B, T and antigen presenting cells, the major histocompatibility complex and is familiar with some of the terms that are applied (eg epitope, immunoglobulin, antigen) in this area of study. The text has, however, been provided with many helpful molecular and cellular 'reminders' and these, together with its open learning and interactive style, will enable most readers to overcome any lack of experience in the key areas.

Progress in the application of biotechnology in the animal sector of agriculture is however being impeded by a lack of sufficient knowledge about the genetics and physiology of animals and by the pressure that arises from a lack of acceptance of this technology in some sections of society. The acceptance of this technology and its application to animal production will have a significant impact on the pace of development in this area. Key to the future development of biotechnologically aided animal production will be questions of safety and ethics. This text does not attempt to dictate about the ethics of manipulating the growth and the reproduction of animals. It is up to the reader to formulate his/her own opinion on such matters. We have, however, included information on matters of safety. The main purpose of the text is to ensure that the reader recognises the benefits that may arise from the application of biotechnology to animal production and understands the routes by which these benefits may be realised. The fears of many regarding the abuse of animals through the application of science and technology are often the result of lack of information or media 'hype'. By providing a description of the present state and the limitations of biotechnological applications to animal production, this text may also provide the reader with the knowledge to respond more fully to the anxieties felt by the 'man in the street' about modern practices in animal production.

Scientific and Course Advisors: Dr M.C.E. van Dam-Mieras
Dr C.K. Leach

Animals in biotechnology - state of the art

Animals in biotechnology - state of the art

1.1 Introduction

Since 1953 when Watson and Crick published their work on the structure of DNA, opportunities to apply molecular biology techniques in the medical, veterinary and agricultural field have increased at a remarkable rate. The comprehensive name given to these developments has been 'biotechnology'. Biotechnology has been defined by the Agriculture Research Service of the U.S. Department of Agriculture as 'the use of living organisms, cells, subcellular organelles, and/or parts of those structures, as well as the molecules to effect physical or chemical changes needed to generate new products for research and commercialisation'.

biotechnology

Despite such broad perspectives, biotechnology is most often simply considered to be synonymous with recombinant DNA techniques or genetic engineering. However, biotechnology in practice typically includes not only genetic engineering, but also some of the older and closely related tools such as cell culture, monoclonal antibodies, bioprocess engineering and the manipulation of reproductive processes. Thus biotechnologists manipulate, not only the genetic make-up of living organisms, but also processes or factors which pre-exist in living organisms but which for practical purposes were previously out of reach. Hitherto the only ways in which these inherent functions could be readily changed were by evolution and selective breeding. Biotechnology offers an extension to effecting change at the organism level by manipulation of cells and individual genes within an organism. It creates a new method of intervention in biological processes with the potential for bringing about changes not otherwise possible.

In subsequent chapters the basic principles and applications of the new science in the veterinary and agricultural field are given. As with any emerging discovery or developing technology, there is often an aura of euphoria associated with the announcement of new findings. In this biotechnology is no different. Even for specialists in the science of biotechnology, it is often very confusing and difficult to distinguish the realistic and viable applications from among the numerous claims and discoveries which are publicised.

major fields of biotechnology in animal production

The state of the art in animal biotechnology is schematically presented in Figure 1.1. It includes three major fields: 1) manipulation of reproductive processes, 2) genetic engineering of macro-organisms and 3) genetic engineering of micro-organisms and molecules including cell-engineering (hybridomas) to produce desired end products such as vaccines, gene probes, monoclonal antibodies and growth promoters.

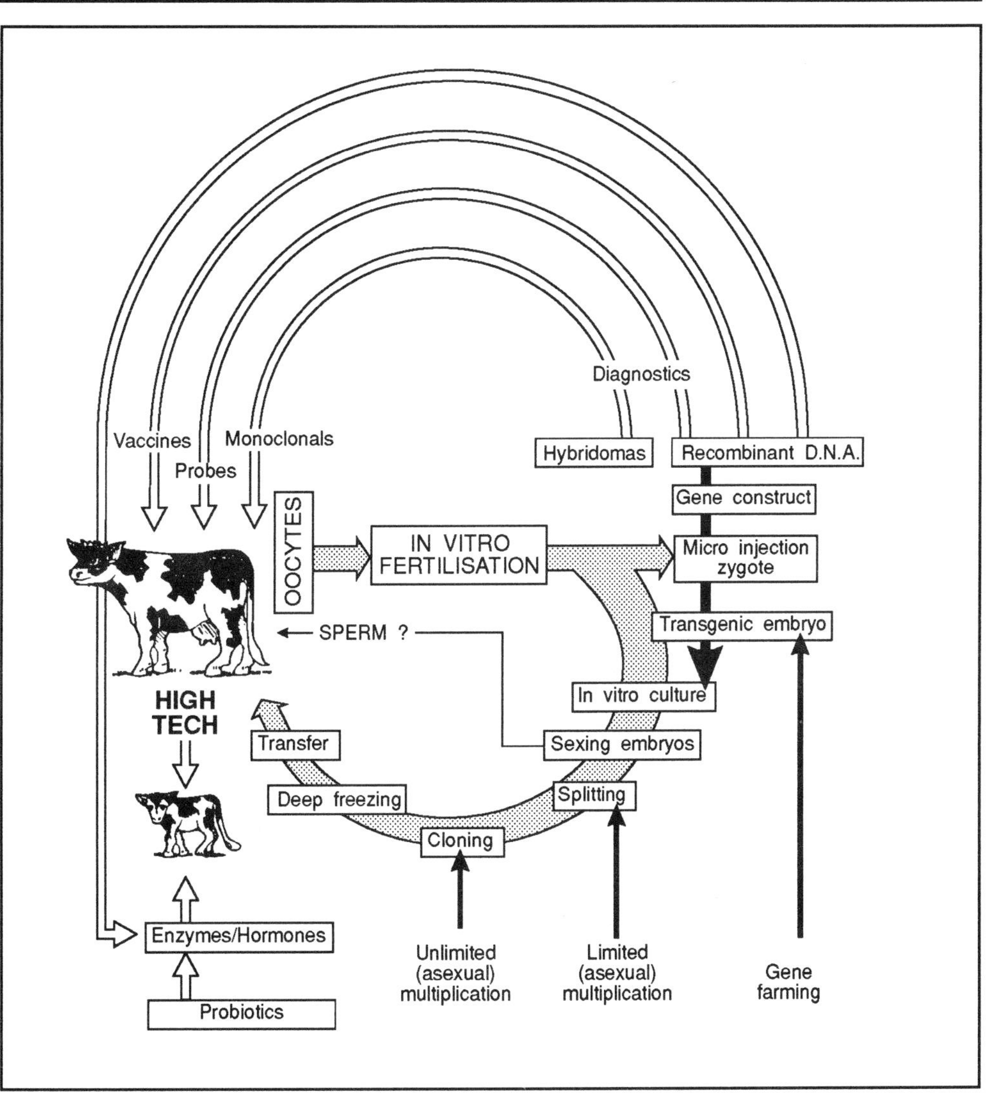

Figure 1.1 A summary of the state of the art in animal biotechnology.

1.2 Manipulation of reproductive processes

artificial insemination

oestrus induction

superovulation

Manipulation of reproductive processes in domestic species started in the 1930s with artificial insemination (AI). Its use became widespread in the 1960s when AI organisations began to make routine use of frozen semen. Manipulation of female reproductive processes started with oestrus induction and synchronisation, initially with steroid hormones and later with prostaglandins. Superovulation techniques with gonadotrophic hormones led to embryo transfer on an increasing scale in the 1980s when embryo freezing techniques became available. AI and embryo transfer techniques offer animal breeders a number of opportunities to enhance the rate of genetic progress in national breeding programmes (Chapter 3).

embryo sexing

embryo splitting

embryo cloning

More advanced techniques for altering reproduction in livestock have been developed in the last decade. These include oocyte removal from genetically superior cows during the dioestrous period, *in vitro* maturation and fertilisation of these oocytes, *in vitro* cultures of fertilised oocytes (zygotes) up to the blastocyst stage, sexing, splitting and cloning of embryos. Splitting and cloning of sheep and bovine embryos became possible after micromanipulation techniques were developed. Marketing of bovine embryos that originate from *in vitro* manipulation procedures such as *in vitro* maturation, fertilisation, culture and sexing, is anticipated in the first half of the 1990s and of cloned embryos in the second half. This is due to the fact that the latter procedure is very sophisticated and involves microsurgery and electrofusion. As soon as these procedures can be used on an industrial level, embryo transfer should partly replace AI in dairy and beef cows.

1.3 Genetic engineering of macro-organisms

Micromanipulation techniques have already been used to produce transgenic mice, sheep, pigs and cows. Gene constructs, coding for a known characteristic, can be injected into the pronucleus of a fertilised ovum. This technique enables the transfer of any gene from any source, no matter how remote the relationship between the donor and recipient is, into a population of animals. The ultimate success rate depends on whether the gene is properly expressed in the recipient and their offspring; ie whether the protein specified by the introduced gene is synthesised or not (Chapter 5). These new techniques fulfil the long-range goal of animal breeders which is to introduce particular genes into the germplasm of domestic animals. This may, together with manipulation of reproductive processes, improve disease resistance, milk production, growth, feed conversion efficiency and overall economic merit. Genetic engineering also yields new insights into basic physiological mechanisms (eg gene regulation), expedites the production of animals with desired production traits and will result in the creation of animals with entirely novel properties (eg secretion of biomedical substances in milk), completely unattainable by conventional breeding and selection techniques. However, it must be stated that considerable research is still needed to improve the actual success rate of 0.5-1% of this technique.

1.4 Genetic engineering of micro-organisms and molecules, including engineering of cells

Vaccines

recombinant DNA and deletion vaccines

The production of recombinant vaccines against various animals diseases that are caused by bacterial, viral or parasitic infections is a field of active research. Most vaccines used in livestock at present are still produced by conventional methods. Only a few genetically engineered vaccines are on the market, such as recombinant DNA vaccines against *E. coli* (pigs), a feline leukaemia vaccine and a deletion vaccine against Aujeszki's disease in pigs.

importance of vaccines

Vaccine development is one of the first and most promising fields of application of biotechnology. At present, an enormous amount of research is going into this area in an effort to develop new, safer and more efficacious vaccines which give the highest protection to the animal economically.

There is still a large number of animal diseases for which protective vaccines are not yet available, especially against parasitic diseases. Some antigens have been virtually impossible to incorporate into vaccines because of the difficulty in growing the micro-organisms or isolating the antigens, or because of their complexity or an inability to adapt them to a viable production method.

Although not many genetically engineered vaccines have been marketed yet, the spin-off of research in this field has led to a better understanding of the nature of microbial immunogens, the mechanisms underlying virulence, the molecular pathogenesis of an infectious disease, and the immunological processes stimulating an animal's immune response.

Research activities are now focused on subunit, recombinant DNA, synthetic peptide, anti-idiotype, deletion mutant, reassortant and vaccinia vectored vaccines (Chapter 7). It is expected that in the next five years an increasing number of genetically engineered vaccines will be marketed.

1.5 DNA/RNA probes

The emergence of recombinant DNA technology has given rise to diagnostic DNA/RNA probes which are based on the ability of single stranded DNA or RNA to form hybrids with a complementary labelled sequence of nucleotides on another strand of DNA/RNA. This means that DNA/RNA sequences unique to a species can be recognised by hybridising, or pairing, test material with DNA, labelled in a recognisable manner. Application of DNA/RNA probes in the diagnosis of infectious diseases is a recent alternative to the established isolation and determination of micro-organisms by cultural and serological methods. Expectations for this new technique are high, because probes may have advantages over the conventional bacteriological and virological methods such as a constant and well defined sensitivity and specificity which enable identification of pathogens in a relatively short time, and their direct detection in clinical specimens. Other probes and techniques are, or are becoming, available for the detection of genetic errors, the determination of sex of embryos, the verification of pedigrees and the monitoring of physiological changes induced by the introduction of new genetic material.

application of probes

Tests to diagnose genetic predisposition of diseases in humans are already leading to screening for early detection of people at high risk for certain diseases, and offer the opportunity to modify environmental factors and thereby alter the course of the disease. At present most clinical diagnoses of infectious diseases have to be confirmed in a laboratory (Chapter 7). However, it may be expected that in the future more and more diagnoses can be confirmed in the course of medical or veterinary practice.

1.6 Monoclonals

Parallel with developments in DNA technology are the advances in immunochemical diagnostic procedures. Serological diagnostic tests, used for many years rely on the detection of antibodies stimulated by invading organisms. The use of antibodies to detect infectious agents, proteins or pharmaceutical substances is exemplified by the expansion of monoclonal antibody technology. This involves the fusion of somatic cells to form a hybrid or hybridoma from both parents. Hybridomas can produce antibodies (monoclonals) specific for a single antigenic determinant (epitope) which can be

monoclonal antibody technology

produced in large quantities. Monoclonal antibodies, when labelled may be used to detect substances in tissues, fluids, or cells even under field conditions.

A practical example is the bovine milk progesterone test for determining the stage of the oestrous cycle (Chapter 3). However, many other monoclonal antibodies have been produced against a number of antigens: tumor cell markers, histocompatibility antigens, lymphocyte differentiation antigens, bloodgroup antigens, bacteria, viruses, fungi, protozoa, helminths, hormones, enzymes, nucleic acids, immunoglobulins, and receptor sites. It is expected that hybridoma technology will deliver more powerful products for both the human and veterinary medical markets in the future.

1.7 Growth promoters

range of hormones produced by genetic engineering

Most of the functions of the body are controlled by hormones, chemical substances produced naturally in different glands, which in minute quantities influence the performance of specialised groups of cells. Individually, they are responsible for the functions of growth, reproduction, milk secretion and the metabolism of the body in general. A number of chemical and pharmaceutical companies have genetically engineered hormones or growth promotants such as human (hST), bovine (bST) and porcine (pST) somatotropin, interferon, lymphokines etc. These are in the field-testing stage, or are ready to market for use in human and veterinary medicine. The use of these products may help to correct growth retardation in children, make hogs grow faster with less fat and more lean production of dairy cattle (Chapter 6).

BST encapsulates many of the questions and fears now being expressed about biotechnology in general. It has led to controversy among dairymen, legislators, and consumer groups. The discussion is centred on moral, public health and economic grounds. We will examine these issues in Chapter 6.

biotechnology contributions to animal diets

Other growth promotants such as antibiotics and non-antibiotic tools including enzymes, yeast cultures, live bacteria and their metabolites and feed pH adjusters are being perfected for use in livestock farming. These products should offer the nutritionists, if used appropriately, great potential for improvement in their animal nutrition programmes. Addition of suitable crude enzyme preparations to diets of both pigs and poultry can lead to improvements in feed conversion efficiency and live weight gains for example. You should also note that biotechnology is also contributing significantly to animal feedstock manufacture and waste treatment.

Entirely new enzymatic processes are being designed for chemical manufacture, waste treatment and biomass conversion and these processes tend to be cheaper, faster and less hazardous, than the ones previously used.

probiotics

Other biological tools are the probiotics which encompasses organisms and substances that contribute to intestinal microbial balance. Applications are mainly in poultry, pigs and calves. The aim is a constant infusion of friendly organisms, such as lactic acid bacteria, via the diet, to prevent colonisation of the gastrointestinal tract by disease-causing organisms. It is thus based on the principle of competitive exclusion.

1.8 Public debate

ethical and regulatory issues

Biotechnological tools for genetic engineering of micro- and macro-organisms have great potential for influencing human and veterinary medicine and animal agriculture in a myriad of positive ways. However, there is great concern for the ethical and regulatory ramifications of this technology. It is important to note that this concern was first raised by scientists -not by regulators- and responsible action by these scientists led to the development of containment guidelines. Concern has now shifted from containment in the laboratory to the release of genetically engineered organisms in the environment. The issue of environmental release of genetically engineered organisms is not simply a national issue, but a global one as well.

There are also some very important ethical issues that arise from biotechnology. They range from the basic question as to whether it is right for us to manipulate creation in any way we choose to specific examples involving the construction, use and release of genetically engineered plants and animals. As far as the genetic manipulation of micro-organisms is concerned, this has largely been accepted. The major public concern focuses onto issues of safety rather than the question of whether it is ethically acceptable to manipulate such systems. The public should, in this context, be re-assured that genetic engineering is conducted within a well regulated and supervised framework. Key to the regulation of genetic engineering within Europe are two EC Directives published in Vol. 33 of the Official Journal of the European Communities L117. These two Directives are the basis upon which EC Member States have brought into force laws, regulations and administrative provisions to ensure that genetic manipulation does not result in personal or communal disasters. As we have implied above, the Directives address two main issues:

EC Directives on genetic manipulation

- the contained use of genetically modified organisms;
- the deliberate release of genetically modified organisms.

These two Directives are of such fundamental importance to both those involved in genetic manipulation and as a re-assurance to the general public as to the safety of genetic engineering, that we have included information on them in Appendix 1. This information is in the form of a commentary on the two EC Directives. You may believe you are familiar with provisions of these two Directives, but it would be well worthwhile checking them out. Note that there are parallel regulations in other parts of the world.

The release of genetically modified micro-organisms, plants and lower animals (insects etc) raises greater concerns over safety then does the release of genetically modified domestic animals. As we will see in later chapters, the manipulation of animal systems and the manufacture of veterinary products such as vaccines involves using genetically modified micro-organisms. Thus the EC Directives are important even to the biotechnological researcher committed to working with mammalian systems.

the intensity of debate about manipulating animals

A major public debate centres on the ethics of experimenting with animals. Much concern arises from the perception of animals suffering in the cause of 'science'. The application of biotechnology to improve animal productivity is seen by some, as an abuse of animals. With the daily advances in our ability to manipulate reproduction (eg *in vitro* fertilisation and embryo splitting) and animal genetics (eg diagnosis of genetic disease, gene therapy), the tension between those opposed to such procedures and those developing them becomes increasingly intense. There are of course those who are opposed to such procedures even when applied to plants and micro-organisms. Their application to animals only serves to heighten the passions of those opposed to the new technology. The production of chimaeric (containing genetic material from two unrelated sources eg sheep and goat) animals also raises important ethical issues and just as much passionate arguments about the rights and wrongs of producing such animals.

Over the next few years, philosophers, theologians and biotechnologists will need to collaborate in discussion and resolution of the ethical questions raised by the radical techniques and products of biotechnology.

Scientists must become more involved in public debate to educate and inform the public of the highly positive roles that biotechnology can and should play in bettering the human and animal condition. A pro-active role by scientists and science managers will lead to a better informed public and will ultimately influence, to a great extent, public perception of the issue and the formulation of regulatory policy. Thus, as you read the remainder of this text, remember that you are not only learning about the technical issues involved in the application of biotechnology to animal production, but that you have a duty to contribute, in an informed way, to the public debate. Within the text, we will predominantly discuss the technical issues involved. But it will be well worthwhile to keep a kind of balance sheet of the advantages that may accrue from the application of biotechnology. Some attention is given to both ethical and regulatory issues.

SAQ 1.1

Can you answer the following questions? If not you really should study Appendix 1 closely.

1) What is a GMMO?

2) What is a GMO?

3) Which of the following techniques for genetically modifying micro-organisms are exempt from the EC Directive on the contained use of genetically modified micro-organisms: cell fusion by artificial means, mutagenesis, somatic animal cell hybridisation, electroporation, self-cloning?

4) What is a Type B operation?

5) To which group should a pathogenic strain of *Salmonella typhi* be assigned (Group I or Group II)?

6) The EC directive on the deliberate release of genetically modified organisms demands that a notification must be submitted. To whom must it be submitted and what should it contain?

Summary and objectives

This chapter provided an overview of the current application of biotechnology to animal productivity. It outlines the main application areas in the manipulation of reproduction, genetic engineering of domestic animals and the manipulation of cellular systems to produce valuable veterinary products. The chapter also raises issues concerning the safety and ethics of applying biotechnology strategies to animal productivity.

Now that you have completed this chapter, you should be able to:

- describe in broad terms the areas of animal productivity which may be improved by the application of biotechnology;
- describe in outline the regulatory framework in which genetic manipulation is carried out;
- explain the terms and conditions of the two EC Directives on genetically modified systems.

Endocrine regulation of the oestrous cycle

Endocrine regulation of the oestrous cycle

2.1 Introduction

Reproduction in domestic animals such as sheep and cattle is always sexual. Male and female gametes meet, fuse and thus create a new organism.

oestrus

Fertilisation is a chance meeting, influenced by the numbers of each type of gamete present, and by the time and the place of their release. In domestic animals fertilisation takes place in the female genital tract. First copulation must take place. The male is capable and willing most of the time, but the female can only mate successfully during fixed periods - oestrus. Internal fertilisation also requires the genital tract to be in an optimal condition. The short life span of both types of gametes, but especially that of the oocyte, may require also that the release of oocytes takes place only when spermatozoa are already present in the genital tract. In those animals in which ovulation occurs as a reflex response to copulation this condition will be met. In large domestic animals ovulation takes place spontaneously during or shortly after the end of oestrus. As spermatozoa need time in the genital tract to complete their final maturation (capacitation) the release of the mature oocyte coincides with the presence of mature spermatozoa in the Fallopian tube.

endocrine system

Successful reproduction occurs when oestrus, the ovulation of the oocyte and the physiological receptiveness of the genital tract are well synchronised. To ensure these prerequisites, a series of glands synthesizes and releases chemical factors (hormones) into the blood which act on target organs elsewhere in the body. Below is a brief description of the endocrine (hormone producing) system. It is written in the form of a brief reminder of the over all structure and operation of the endocrine system to provide a framework in which to understand the operation of the hormones which regulate the oestrous cycle. The assumption has been made within this chapter that you are familiar with the main features of the endocrine system. If, on reading the next paragraph you realise that your knowledge of the endocrine system is too elementary, we have provided a fuller account of this system in Appendix 2. This Appendix has been designed to ensure you have the appropriate background in endocrinology to cope with this and the subsequent chapters.

The endocrine system includes amongst other glands a) the hypothalamus, b) the pituitary gland or hypophysis, c) the ovaries, and d) the uterus. These organs and glands are responsible for the hormonal regulation of the oestrous cycle, and indeed for the whole process of reproduction. Use Figure 2.1 to help you follow the description of the hormones and regulators involved in the regulation of the oestrous cycle.

The endocrine system is not autonomous. It is dependent on the central nervous system and especially the hypothalamus. The latter forms part of both the central nervous and endocrine systems. At the level of the hypothalamus the two systems have a neuro-endocrine relationship, whereby nervous (electrical) signals are converted into hormonal (chemical) signals. Through this link the endocrine system receives information from the body of the animal (the internal milieu) and also from the external environment. Stimuli from both these internal and external sources are translated via the cells of the hypothalamus. In turn they result in the synthesis and secretion of releasing and inhibiting factors (or hormones) and the hormone oxytocin, and also in oestrous behaviour. Oxytocin is released into the blood in the neuro hypophysis or posterior pituitary and acts directly on peripheral target organs (the uterus and mammary gland) causing smooth muscle to contract and is not specifically involved in the regulation of the oestrous cycle. The releasing and inhibiting factors reach the adenohypophysis or anterior pituitary via the blood. There they induce or inhibit the synthesis and release of the gonadotrophic hormones in the different cell types of this gland. The gonadotrophic hormones are LH (luteinising hormone), FSH (follicle stimulating hormone) and prolactin.

neuro-endocrine

releasing and inhibiting factors

oxytocin

gonadotrophic hormones

The release of LH and FSH is controlled by GnRH (gonadotrophin releasing hormone) and that of prolactin is controlled by an inhibiting factor (PIF) called dopamine. The three protein hormones LH, FSH and prolactin stimulate the ovary to produce and release mature gametes, and to synthesise and release the protein hormone inhibin and the sex hormones: oestrogens, progesterone and androgens. The last three groups of hormones, which chemically are steroids, influence the genital tract so that fertilisation of the oocyte can take place. If fertilisation occurs and the animal becomes pregnant, oestrus and ovulation cease for a long time. If the animal does not become pregnant, the genital tract and particularly the uterus secretes the local mediator (local hormone) prostaglandin-F2α (PGF2α). The latter acts on the corpus luteum in the ovary (formed following ovulation), induces its regression, stops the progesterone synthesis and opens the way for a new oestrus. The ovarian hormones secreted into the blood stream influence not only the genital tract but also the hypothalamus. They form part of the internal milieu and by feedback complete the control loop shown in Figure 2.1.

GnRH

dopamine

sex hormones

fertilisation

$PGF_{2\alpha}$

In the next section the regulation of the oestrous cycle will be discussed step by step on the basis of this outline control scheme.

SAQ 2.1

What are the three major types of hormone that regulate the reproductive system? Name the sources of these hormones and their principal target organs.

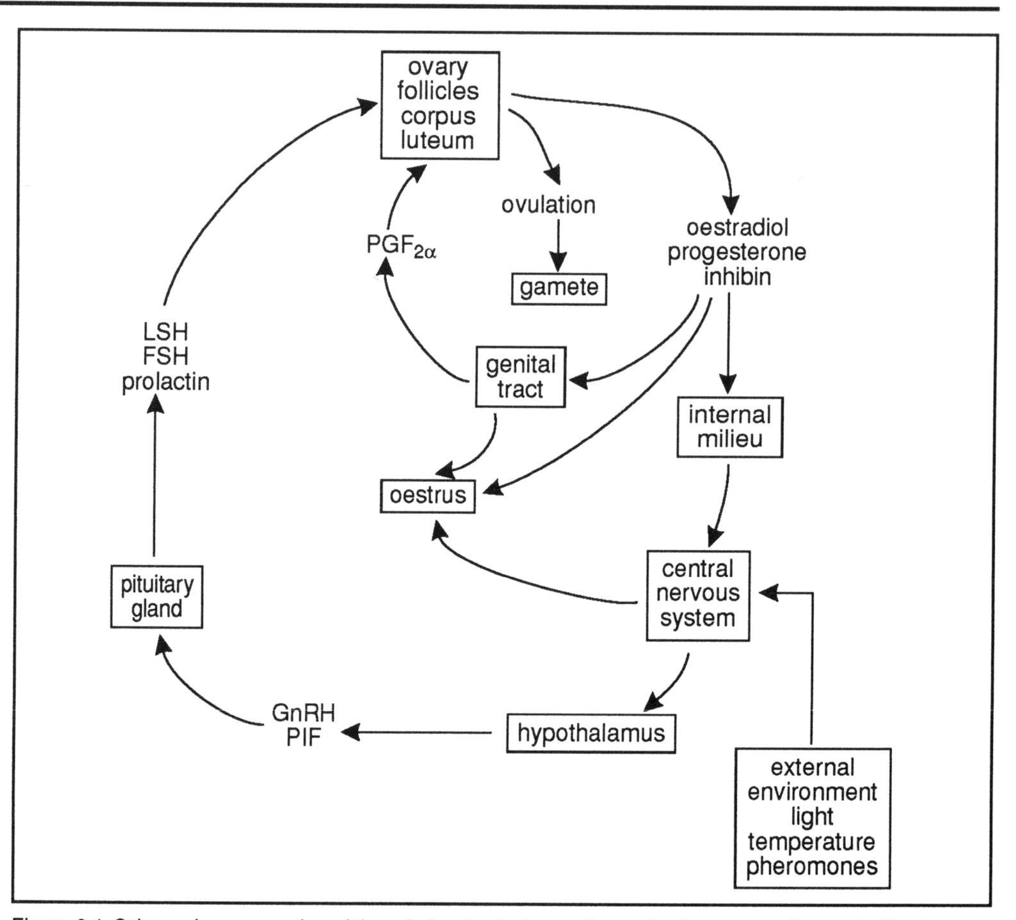

Figure 2.1 Schematic presentation of the relationship between the endocrine system (internal milieu) and the external environment in the regulation of the oestrous cycle in domestic animals (sheep and cattle). (Note that only the major components are shown).

2.2 Oestrus

arcuate nucleus

Oestrus is the psycho-physiological period in which the female is willing to mate with the male. Groups of neurones (nuclei or sex-centres), for example the arcuate nucleus (Figure 2.2) in the hypothalamus, are involved in the regulation of sexual activity and the activity of the sex organs. Thus the arcuate nucleus is a major source of GnRH.

Π Use Figure 2.2 to become familiar with the structure and the naming of parts of the hypothalamus, pituitary and associated structures. You should for example be able to name the connecting part between the hypothalamus and the pituitary.

environmental cues

At sexual maturity this centre will be influenced by steroid hormones leading to continual sexual activity in the male and cyclical activity in the female. Hypothalamic nuclei are also influenced by environmental cues such as light, temperature and pheromones which can affect libido and fertility. Besides these nervous influences, the sex-centre receives information through humoral (blood) routes. Changes in the concentrations of hormones and nutrients and in blood temperature are directly registered by the nerve cells in the hypothalamus.

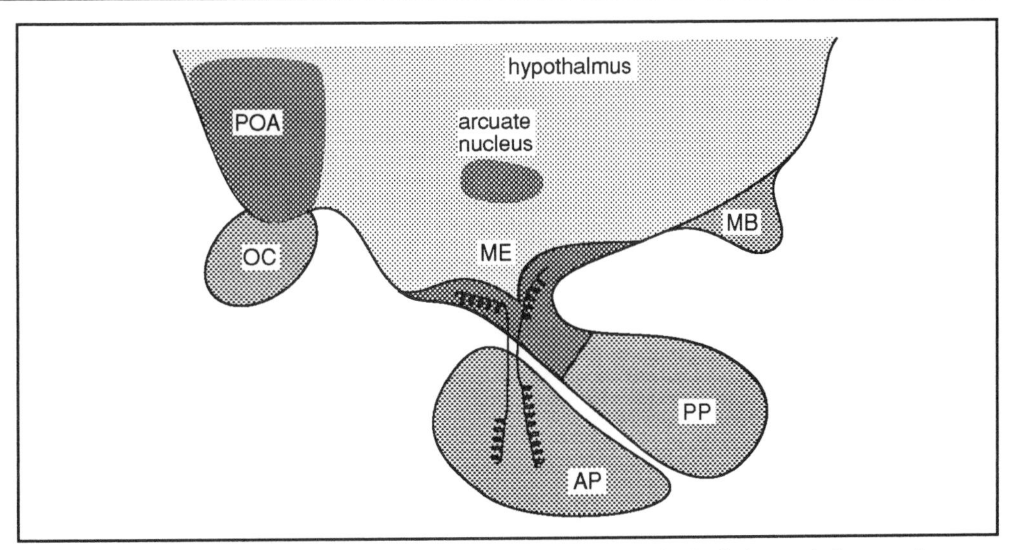

Figure 2.2 Distribution of GnRH in the hypothalamus. The more densely shaded areas indicate regions where the releasing hormone has been located. POA, pre-coptic area; ME, median eminence; OC, optic chiasma; MB, mamilliary body; AP, anterior pituitary; PP, posterior pituitary.

Food intake and illness can also influence reproduction by their effects on the neurosecretory cells of the sex-centre. Stimulation of these neurosecretory cells, causes them to release their secretions (the neurosecretion) into the bloodcapillaries in the median eminence (Figure 2.3). Via the blood stream the neurosecretory products reach the adenohypophysis and so stimulate the secretion of LH by this organ.

Π In Figure 2.3 we have numbered neurones 1-5. Which neurone causes the release of releasing hormone which is transported via the tanacyte to the portal vessel? Which neurones are depicted as non-secretory neurones? Which secretory neurones cause releasing hormone levels in the portal vessels to rise?

Your examination of Figure 2.3 should have revealed that the releasing hormone product of neurone 2 is transported via the tanacyte. The non-secretory neurones are 3 and 4, whilst secretory neurones 1, 2 and 5 produce releasing hormones which will cause the levels of these hormones to rise in the portal vessel. Figure 2.3 shows, therefore several routes by which neurosignals can result in a rise in releasing hormone levels in the portal vessel.

LH-peak release

The oestrus is marked endocrinologically by a very high concentration of LH in the peripheral blood. This so-called LH-peak release at the onset of oestrus induces ovulation at the end or shortly after the end of oestrus.

oestradiol

Oestrus and ovulation result from the interactions of internal factors and external cues. These interactions take place in the hypothalamus. It is still difficult to determine which factors play a role in the induction of oestrus, LH-peak release and ovulation. It is however generally accepted that oestradiol-17β (E2) has a dominant position. Injection of oestradiol-17β into ovariectomised animals causes oestrous behaviour (Table 2.1). The duration of this oestrous phenomenon is dependent on dose and whether or not the animal is pretreated with progesterone. Injection of oestradiol-17β also results in a peak-release of LH.

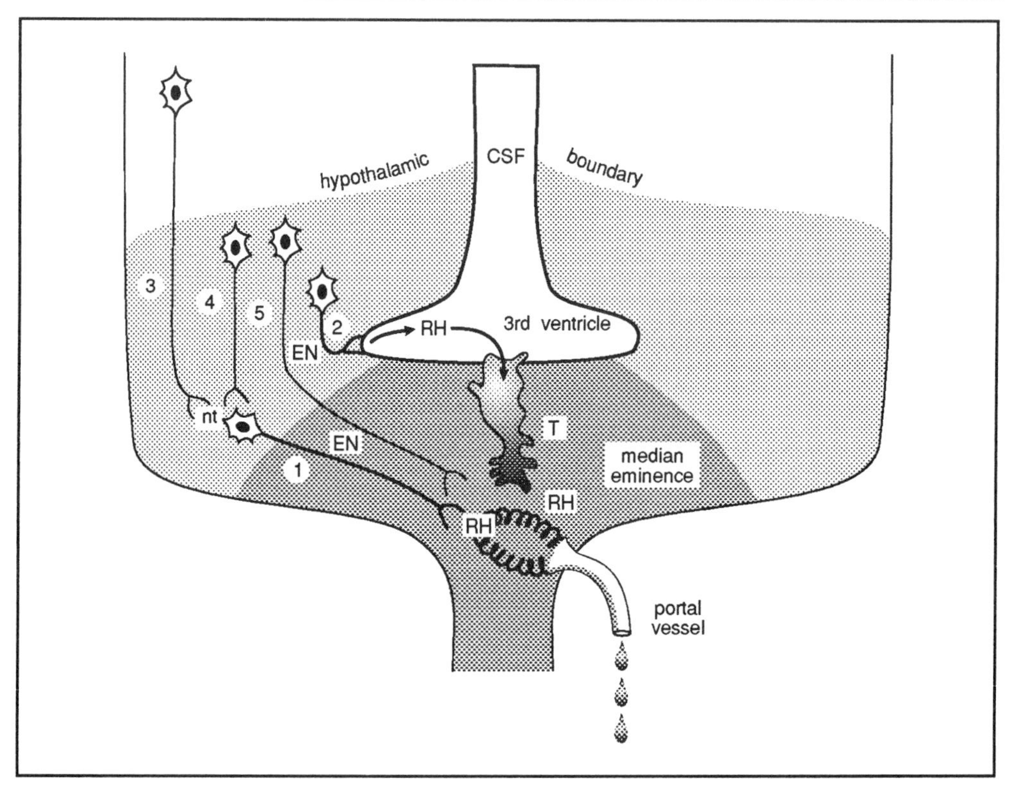

Figure 2.3 Diagram of proposed routes of transport of releasing hormones (RH) from their sites of production to the portal vessels, and the different arrangements by which conventional neurones communicate with endocrine neurones. EN, endocrine neurone; nt, neurotransmitter; T, tanacyte; CSF, cerebrospinal fluid.

Dose oestradiol (mg)	LH release (ng ml^{-1})	Time interval between injection and LH-peak release (hours)	Duration oestrus (days)	Time interval between onset of oestrus/LH-peak release (hours)
10.0	23 ± 1.7	18	5	4
2.5	18 ± 1.3	16	1.5	4
1.0	24 ± 1.8	15	1	3
0.5	19 ± 1.4	16	-	3

Table 2.1 Relationship between dose of oestradiolmonobenzoate injected into ovariectomised cows and the level of LH in the blood and the duration of oestrus.

The data in Table 2.1 showed a relationship between LH-peak release and oestrous behaviour. We might ask are they obligatorily linked or is there evidence that they are independently controlled?

silent heat

We know from Table 2.1 that a hormone can have simultaneously an effect on the nervous system (oestrous behaviour) and on the endocrine system (LH peak release). However, to induce oestrus more oestradiol-17β seems to be needed than is needed to procure LH-release (see Table 2.1: 0.5mg oestradiol injected). Experiments thus show that under certain circumstances the concentration of oestradiol-17β in the blood is too low to induce oestrus but high enough to obtain LH-release and ovulation. LH-release and ovulation without oestrous behaviour is called a silent heat.

The phenomenon of silent heat under these circumstances is dependent on pre-exposure to progesterone. This is always the case during the normal oestrous cycle as is shown in Figure 2.4. At first glance this figure looks very confusing. It does, however, summarise a lot of information, so let us see if we can sort it out. Use Figure 2.4 to follow the description below.

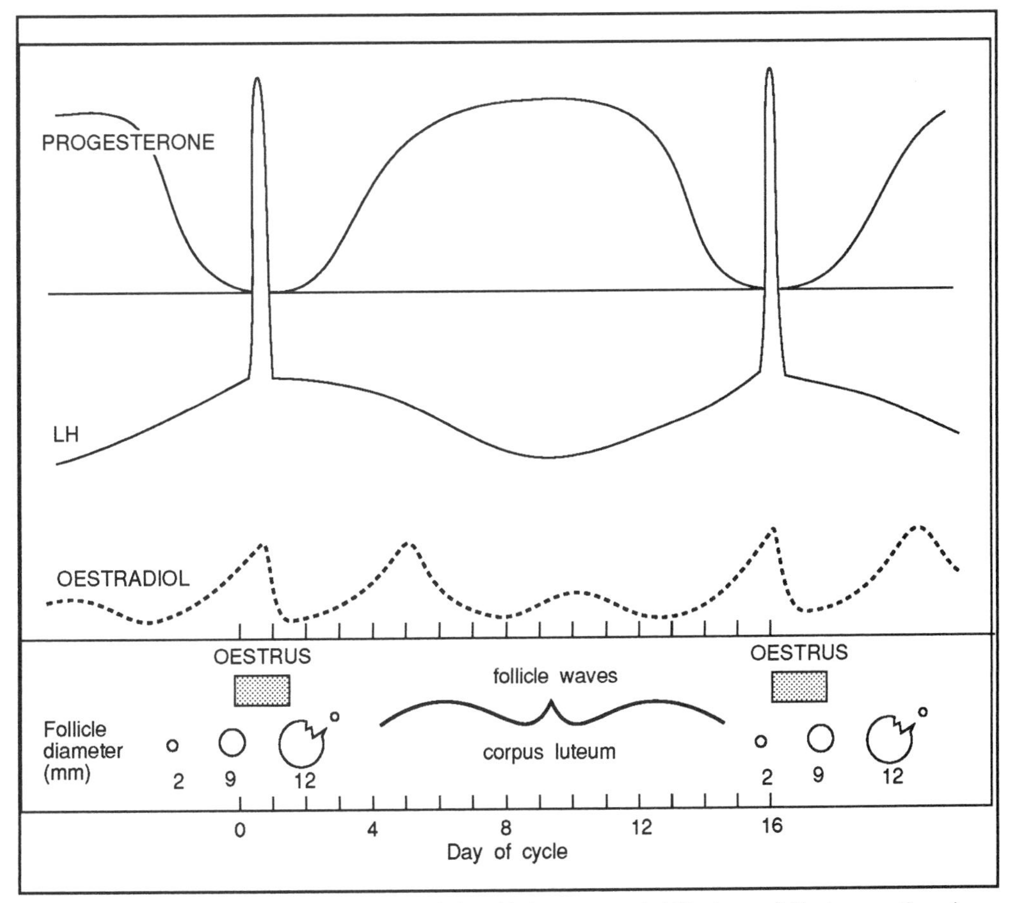

Figure 2.4 Schematic representation of the relationship between tonic-LH-release, follicular growth and oestrogen synthesis, LH peak release and ovulation and the development and regression of the corpus luteum during the oestrous cycle in sheep. (see text for further details).

The bottom axis shows the time scale of the cycle. (How long is the cycle depicted in the figure?) The next sequence represents the development of the follicle(s) within the ovary. Note that in sheep each follicle is about 2mm in diameter. As oestrus approaches, the follicle swells to about 9mm leading to ovulation about 2 days after the onset of oestrus. The next graph shows how the levels of oestradiol fluctuate during the cycle.

Oestradiol levels are highest at oestrus but fluctuate in between successive periods of oestrus. Note that the luteinising hormone (LH) levels are very high early in oestrus, but rapidly fall away again. Progesterone levels fall as oestrus approaches, but rapidly rise again after oestrus if pregnancy does not take place.

Thus as the progesterone level falls prior to oestrus, the oestradiol levels rise and peak amounts of LH are released. The high levels of oestradiol and LH and the low levels of progesterone are typical of oestrus. If oestrus and subsequent ovulation does not result in pregnancy, progesterone levels rise, LH levels fall and oestradiol levels fall. Towards the end of the cycle, the levels of progesterone fall and the cycle is ready to commence again.

If we are to manipulate reproduction, then one strategy is to modify the frequency of oestrus. This, of course, will involve influencing the hormone status. We will examine this strategy in a later section so it is important that you remember the overall scheme described in Figure 2.4.

Π It would be quite a good test to draw out your own scheme of the changes in hormone levels during the oestrous cycle and then check it against Figure 2.4.

2.3 Oestrous cycle

oestrous cycle

As we have seen, the oestrous cycle is the periodic appearance of oestrous behaviour with a species specific regularity. In domestic large animals, the interval between two oestrous periods is not influenced by mating. In sheep this interval is around 17 days, in the cow 21 days. The first day of the cycle ie the day of oestrus is called D-0. The following days are designated D-1 etc until the next first oestrus (Figure 2.4). Some animal specific differences are summarised in Table 2.2.

Animal	Onset of puberty	Cycle length (days)	Oestrus duration (hours/days)	Time of ovulation
Horse	12-18 months	21 days	5 days	24-48h before end of oestrus
Cow	8-12 months	21 days	18 h	30 h after onset of oestrus
Pig	5-7 months	21 days	1.5-3 days	second part of oestrus
Sheep	6-16 months	17 days	24-36 h	24-30 h after onset of oestrus

Table 2.2 Onset of maturity and oestrous cycle duration.

2.4 Oestrous and anoestrous seasons

oestrous and anoestrous seasons

In the sheep oestrous behaviour does not appear over the whole year. The period of oestrous activity is called the oestrous season, the period of rest from sexual activity is called the anoestrous season. In the northern hemisphere and in particular at latitudes of about 52o the oestrous season starts in September and ends in February. The rest of the year is the anoestrous season. In the southern hemisphere the pattern is reversed. In both hemispheres the oestrous season coincides with the period of short day-length and with decreasing daily temperature.

day-length

The sheep is therefore an animal that comes in oestrus when days are shortening. The horse also has an oestrous season but this starts when day-length is increasing. The way external factors like day-length (or photoperiod) and temperature act on oestrous behaviour has probably resulted from natural selection. Young animals have the best chance of survival when they are born in a period with enough natural food. Sheep with a pregnancy length of nearly five months come into oestrus between September and February. Horses with a pregnancy of 11 months should come in heat in the period between February and September. In both cases the young will be born in the spring. The effects of day-length on the hypothalamo-pituitary-ovarian axis are summarised in Figure 2.5.

photoperiod

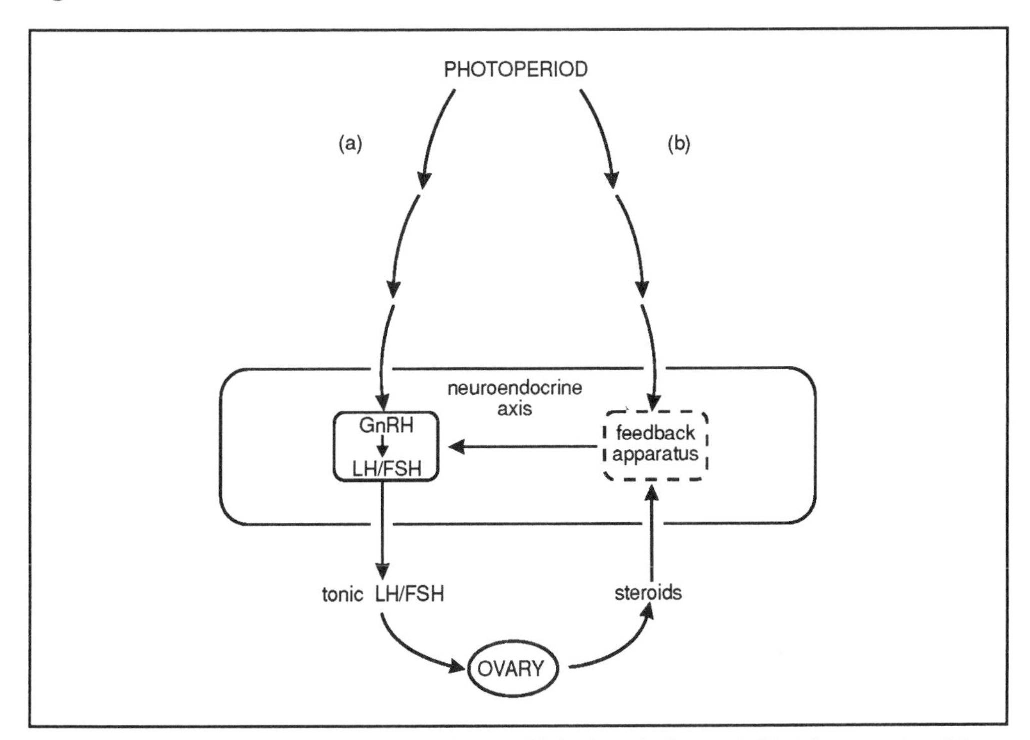

Figure 2.5 Schematic representation of the influence of light through photoperiod (environmental cue) (a + b) and steroids (internal milieu) on the secretory activity of neurones in the hypothalamus (neuroendocrine axis) leading to release of GnRH then LH and FSH.

Π Using the information concerning the occurence of anoestrous seasons and the effect of day length on oestrous behaviour, write down a strategy for increasing reproduction rates in these animals.

The answer we anticipate that you would write was that if the day-length (and perhaps temperature) the animals were exposed to were controlled (ie by keeping them in artificial environments) their 'biological clocks' might be fooled to think that they were in their normal oestrous season and thus would enter oestrus.

SAQ 2.2

A herbivore has a gestation period (time from fertilisation until the birth of the young) of 18 months

1) At what stage of the year would you anticipate the animal's oestrous season to be if the animal lived in the northern hemisphere around latitude 50-55°? Give reasons for your answer.

2) How is the oestrous season linked to a particular phase of the year?

epiphysis or pineal organ

Recently much attention has been paid to the role and function of the epiphysis or pineal organ, a small organ in the roof of the diencephalon of the brain. In its early evolutionary history, this organ consisted of light sensitive (photo-receptive) cells and secretory cells. In mammals the day-length registered by the pineal gland via the eyes and complex neural connections influences synthesis and the secretion of several indoleamines. The most important of these is melatonin. This pineal hormone seems to act on the nuclei in the hypothalamus to modify the input of neurotransmitters in the neurosecretory cells which in turn affects GnRH secretion. Thus day-length can influence the output of GnRH by neuro-secretory cells. In the summer LH release in the sheep is reduced. However, little is known about the precise regulation of GnRH release.

melatonin

negative feedback

positive feedback

Two possible interlinked systems have been proposed. You should note that LH release is regulated in two ways. Firstly there is a sharp peak release of LH during oestrus (Figure 2.4). There is also a low level of LH release (so called continuous, tonic LH release) over the whole period of the oestrous cycle. However, tonic LH release tends to rise in the second half of the cycle as progesterone levels fall. One system controls the LH-peak release (surge centre; Figure 2.6), the other, the continuous tonic LH release (episodic or tonic centre). The latter is sensitive to negative feedback by steroids.

Progesterone always has a negative feedback effect on the tonic centre, but oestradiol-17β only does so when the days are lengthening. The negative feedback by oestradiol-17β declines when the day-length is decreasing. The cyclic LH-peak release is dominated by a stimulatory effect of oestradiol-17β through positive feedback on the surge centre. This means that under the influence of oestradiol-17β during the oestrous cycle, the number of receptors for GnRH in the pituitary gland is increasing. When the concentration of oestradiol-17β in the blood increases the pituitary cells respond by releasing suddenly large amounts of LH (Figures 2.4 and 2.5). One should realise that steroid synthesis in the ovary, in turn, is under the control of LH. The lower the LH release the less steroids (especially oestradiol-17β) are secreted by the ovary. When the days are lengthening the pituitary cells are inhibited by oestradiol-17β (negative feedback). This leads to a decrease of LH release until the moment that the oestradiol-17β synthesis is so low that neither LH-peak release and ovulation nor oestrous behaviour results (anoestrous season). When the days are shortening, more and more LH will be released resulting in more oestradiol-17β secretion by the ovary. Then follows as a matter of course the first LH-peak and ovulation to mark the beginning of the oestrous season.

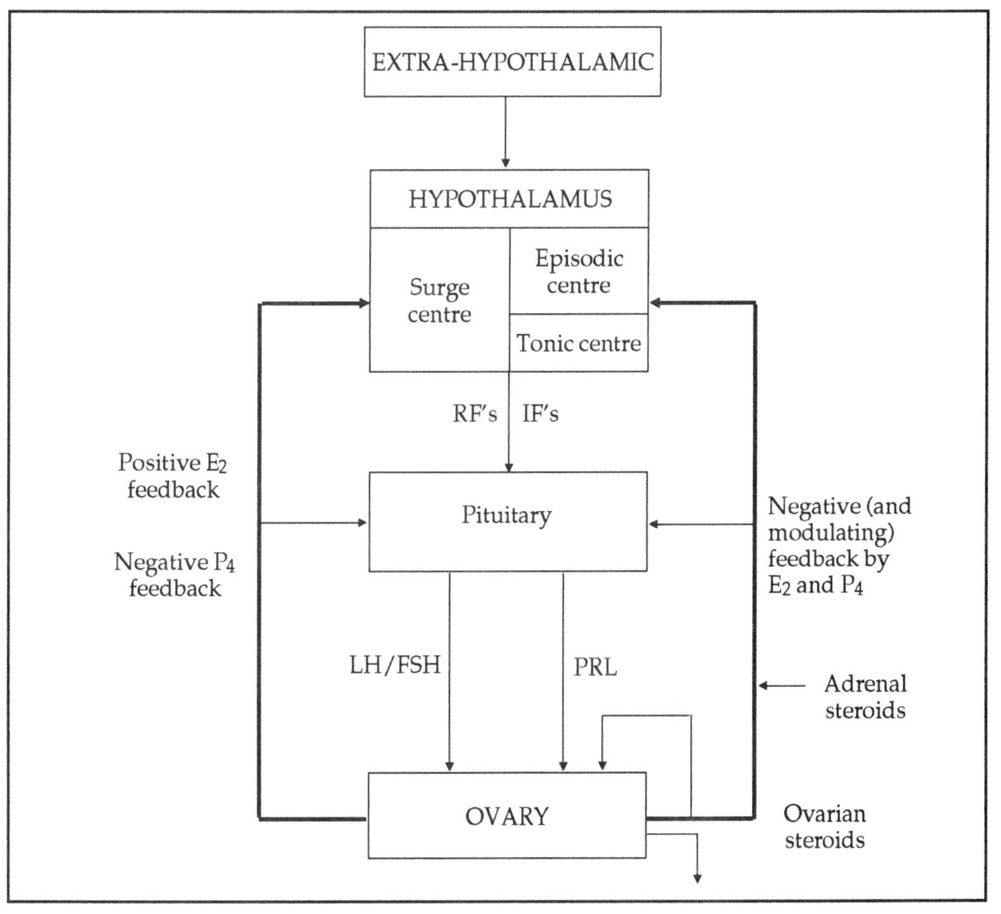

Figure 2.6 A schematic diagram representing the endocrine control of the brain-pituitary-ovarian axis in the mature female mammal. E_2, oestradiol-17β; P_4 progesterone; PRL prolactin. LH = luteinising hormone; FSH = follicle stimulating hormone; RF = releasing factor; IF = inhibiting factor.

silent heat

Since no pre-exposure to progesterone pretreatment had occurred during the anoestrous season, the first LH-peak and ovulation are not accompanied by oestrous behaviour. Thus the oestrous season starts with a 'silent heat', that is an ovulation without overt signs of oestrus.

Species' differences in the oestrous behaviour during the year can be explained by differences in the sensitivities of GnRH producing hypothalamus cells to exogenous and endogenous factors.

dioestrus

The period of time between two oestrous periods is called dioestrus. The rhythm of oestrus-dioestrus on the one hand and of oestrous and anoestrous seasons on the other is also the result of changes in the activity of endocrine organs. These changes can be detected by measuring the concentrations of hormones in the blood.

SAQ 2.3

Which of the two main ovarian steroids are responsible for the regulation of a) cyclicity of oestrus, and b) the rhythm of oestrus and dioestrous seasons?

2.5 Hormones from the hypothalamus

2.5.1 Releasing and inhibiting factors

neurosecretory nuclei

In the hypothalamus groups of nerve cells or neurosecretory nuclei secrete their products into blood vessels in the median eminence or in the neurohypophysis (Figure 2.7). These releasing and inhibiting factors are secreted into the portal vascular system of the median eminence. The former are small peptides and they stimulate the synthesis and release of the gonadotrophic hormones LH and FSH by the hypophysis. The latter are neurotransmitters like dopamine which inhibits the release of prolactin.

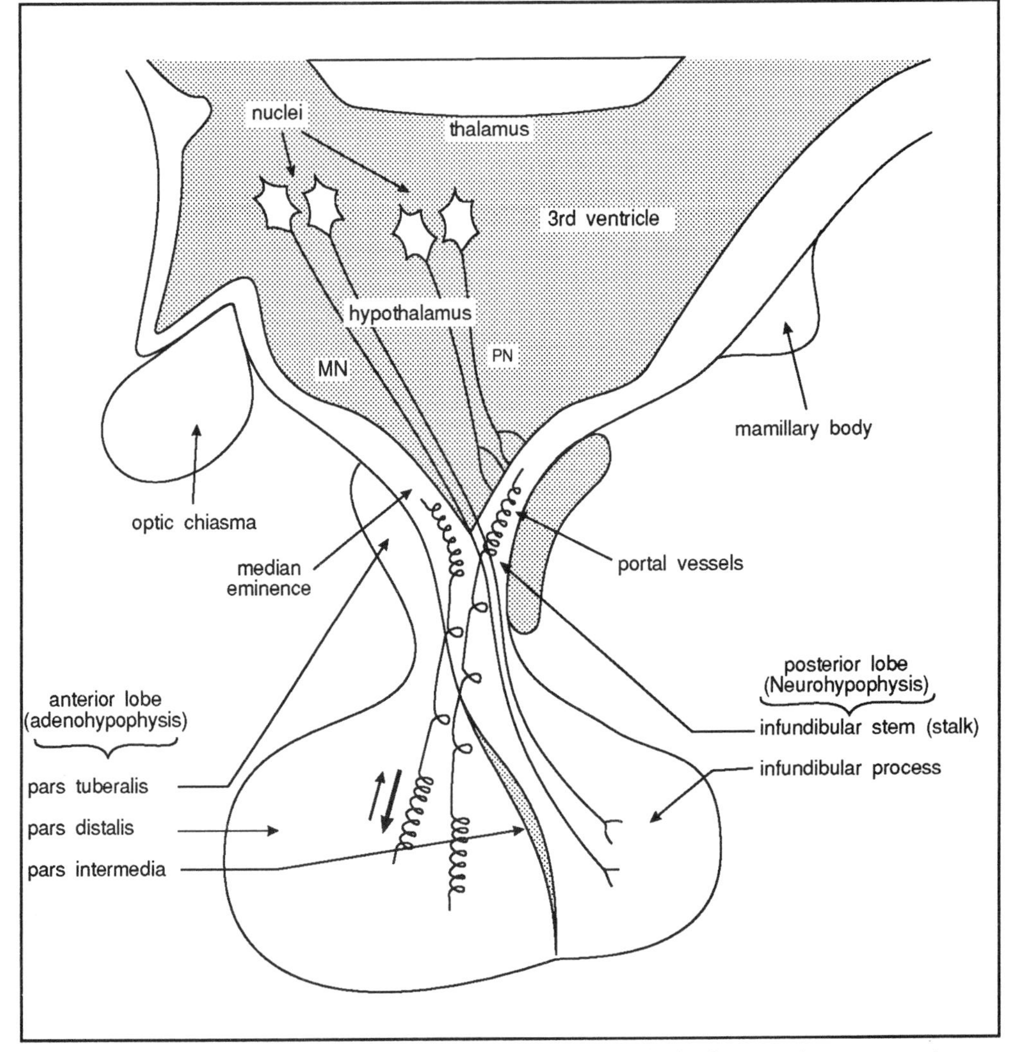

Figure 2.7 Schematic presentation of the direct and indirect communication between the neurosecretory nuclei in the hypothalamus and the neuro- and adenohypophysis respectively.

Π Figure 2.7 looks fairly complex, so examine it carefully. It represents the communication between the neurosecretory nuclei in the hypothalamus and the anterior and posterior lobes of the hypophysis (adenohypophysis and neuropophysis). Start with the nuclei of the neurosecretory cells at the top. Follow their axons down. You will see that some end adjacent to the portal vessel. Others pass all the way down to the hypothesis. From Figure 2.7 answer the following: 1) Which lobe of the hypophysis receives neurosecretory hormones from neurones in the hypothalamus via the portal vessels? 2) How many parts of the anterior lobe of the hypophysis are labelled, and what are they called?

You should have spotted that the anterior lobe receives neurosecretory hormones via the portal vessels. The three parts of the anterior lobe we have labelled are the pars tuberalis, pars distalis and pars intermedia.

Of particular interest here is the regulation of LH and progesterone. Use Figure 2.8 to help you follow this description. LH has a strong positive influence on the production of progesterone by the ovaries. As the progesterone levels in the blood rises, it is detected by nerve cells in the hypothalamus. These stimulated nerve cells reduce their production of releasing hormone (GnRH) which in turn reduces the production of gonadotrophic hormone (LH) by the adenohypophysis.

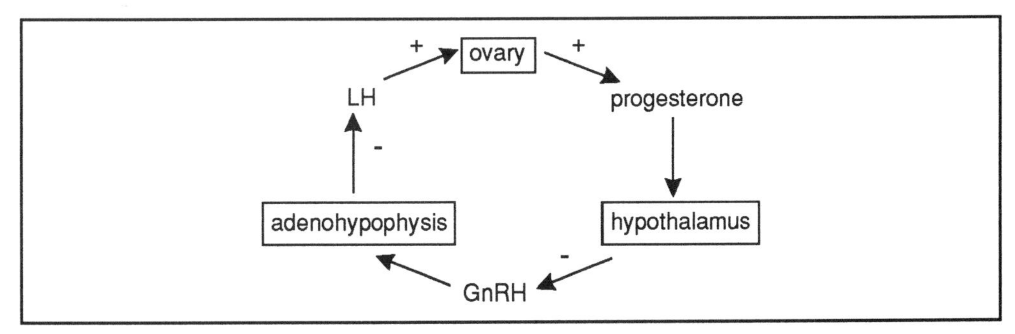

Figure 2.8 An example of negative feedback loop between a steroid hormone and a gonadotrophic hormone.

The feedback system acts as a kind of safety control. This mechanism is not restricted to the hypothalamo-hypophyseal-ovarian relationship. Hypophyseal hormones can also act directly on their own secretion by influencing the release of their releasing hormones (short loops). However, the release of GnRH is strongly controlled by steroids, and the release of pituitary FSH by the ovarian hormone inhibin and the release of prolactin by dopamine.

Π It would be helpful for you to draw these feedback loops out onto a sheet of paper as a form of revision. Alternatively copy out Figure 2.8 onto a larger piece of paper and add these other control loops to it.

2.5.2 Hormones from the neurohypophysis

oxytocin

Two hormones are directly released into the blood in the neurohypophysis, though both originate in the hypothalamus. One is called vasopressin and regulates the water reabsorption in the kidney. The other is oxytocin. The latter plays a role in reproduction. It acts on muscle and myoepithelial cells leading to contractions of the myometrium of the uterus during parturition and to ejection of the milk in response to suckling (Figure 2.9).

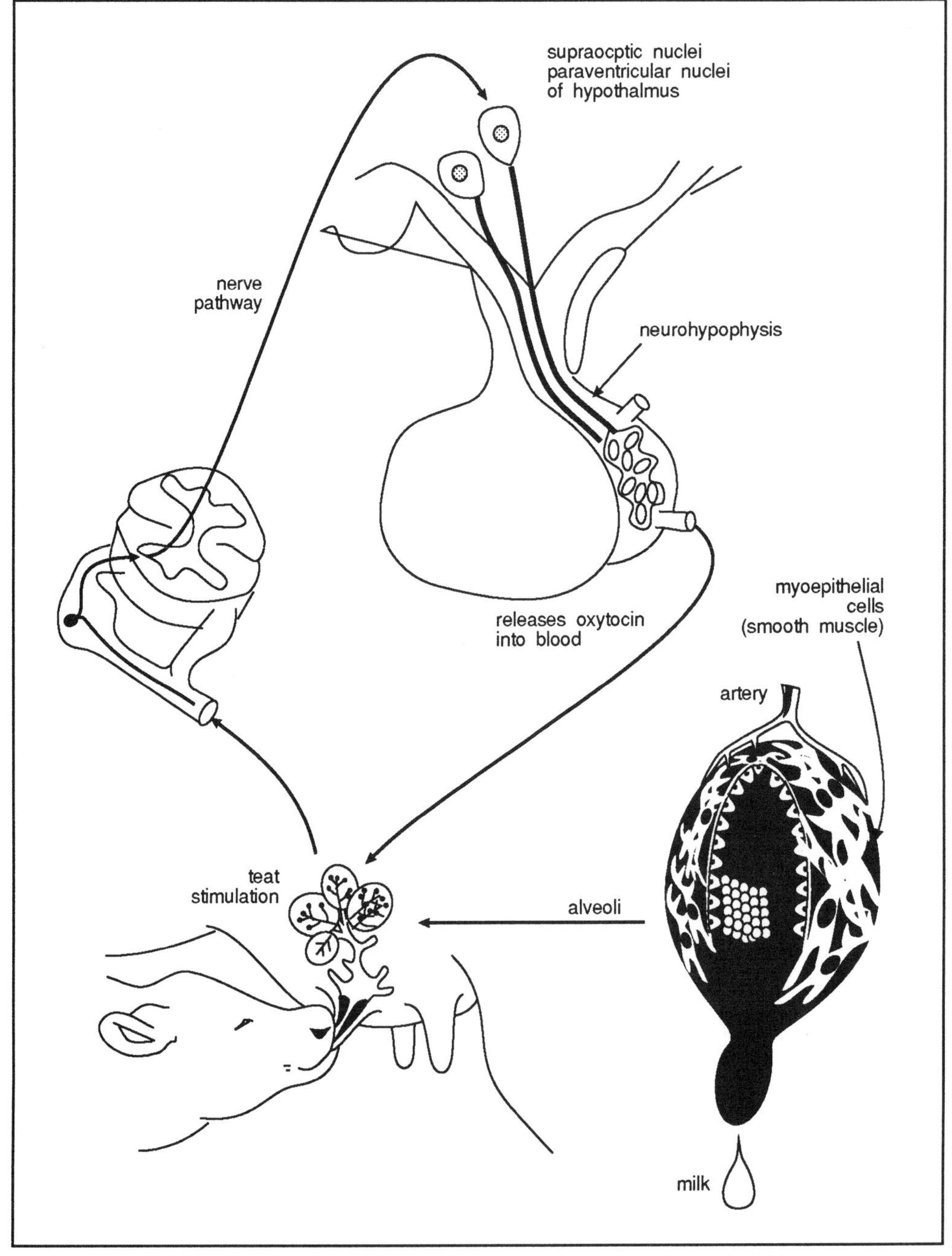

Figure 2.9 Milk let-down may be considered a neuroendocrine reflex. The stimulation of the teat induces a neural signal to the hypothalamus to release oxytocin from the neurohypophysis, which stimulates the myoepithelial cells to constrict the alveoli resulting in milk secretion.

SAQ 2.4

How does the link between the hypothalamic nuclei and oxytocin release from the neurohypophysis differ from the release of releasing and inhibiting hormones by hypothalamus nuclei?

2.5.3 Hormones from the adenohypophysis

LH, FSH and prolactin

The pituitary produces among other things the gonadotrophic hormones LH, FSH and prolactin. They regulate the activity of the gonads (Figure 2.1). Prolactin is a protein consisting of 198 amino acids, LH and FSH are glycoproteins. Protein hormones cannot enter the cells of their target organs. These cells contain in the outer membrane special molecular configurations (receptors) to which the protein hormones can bind. Bound to the receptor, the hormone activates the enzyme adenylate cyclase, by which process intracellular ATP becomes dephosphorylated and cyclic AMP is formed. Cyclic AMP is a second messenger and induces cell specific reactions (Figure 2.10), which in the example given results in the synthesis and release of progesterone. The role of secondary messenger, especially cyclic AMP, are further discussed in Appendix 2. If you are unfamiliar with the principles involved, we would recommend you read the relevant section of the appendix before reading on.

second messenger

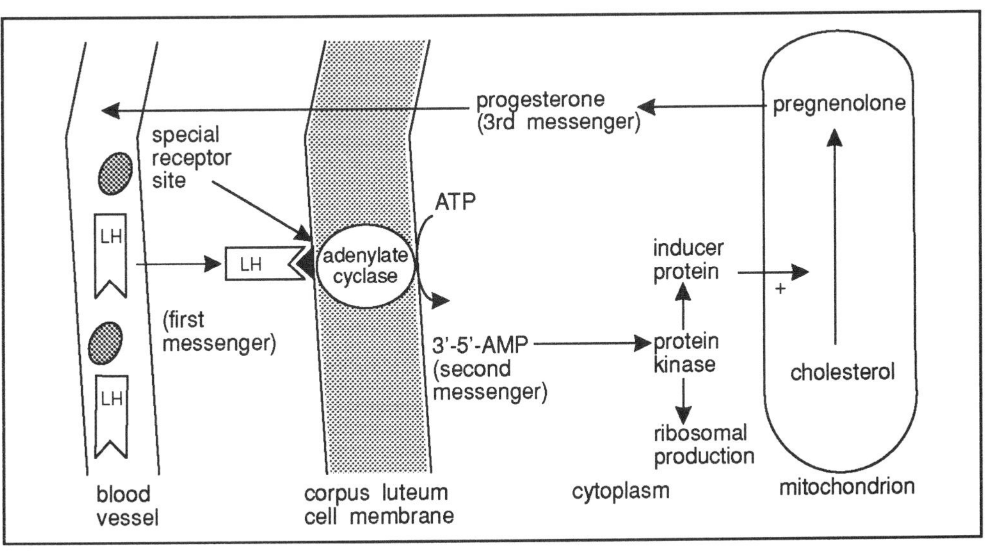

Figure 2.10 Mechanism of action of a protein hormone.

Follow the sequence in Figure 2.10 carefully, you will see that LH stimulates the cells of the corpus luteum to secrete progesterone.

Π In Figure 2.10 the protein hormone is identified as LH. What aspects of the model would be different in the case of another protein hormone? (List two or three differences).

Obviously the hormone itself shown in the bloodstream would be different. The molecular configuration of the receptor would be specific to the hormone to bind on the basis of a lock-key interaction. The roles of adenylate cyclase and cAMP would be the same but the cell's overall response would be specific to the particular protein hormone.

Some specific biological functions of the three gonadotrophic hormones are:

LH (luteinising hormone):

- synthesis of steroids;
- pre-ovulatory growth of the graafian follicle;
- granulosa cell differentiation;
- oocyte maturation;
- ovulation;
- support of the corpus luteum;

FSH (follicle stimulating hormone):

- follicular growth;
- conversion of androgens into oestrogens;
- oocyte maturation;

Prolactin (luteotrophic hormone (LTH) in sheep):

- support of the corpus luteum;
- lactation.

These functions of the three gonadotrophic hormones will be discussed in Section 2.6.

2.5.4 Alternative sources of gonadotrophic hormones

hCG (human chorionic gonadotrophin)

For the manipulation of the oestrous cycle and for clinical use in daily practice, medicine relies on gonadotrophic hormones. Since the concentration of these hormones in the pituitary is relatively low, other sources are used for their isolation and purification. hCG (human chorionic gonadotrophin) that acts like LH is isolated from the urine of a pregnant woman.

eCG (equine chorionic gonadotrophin)

eCG (equine chorionic gonadotrophin) that acts like both LH and FSH is isolated from the serum of pregnant mares and is also called PMSG (pregnant mare's serum gonadotrophin).

Placental-lactogen (or somato-mammotrophin) is produced by the trophectoderm cells of an embryo after attachment. It is also isolated from the serum of a pregnant animal and acts like prolactin and/or growth hormone.

Now that we have learnt a lot about the hormones involved in the regulation of the oestrous cycle, let us examine what happens in the ovary.

2.6 The ovary

2.6.1 Growth and degeneration of follicles

corpus luteum

The macroscopically observable changes on and in the ovary are the number and the size of follicles on the surface of the ovary as well as the growth and the regression of the corpus luteum (CL).

The visible follicles all belong to the tertiary follicular stage (Figure 2.11). Note that we have labelled parts of the tertiary follicles using their technical names.

Draw a stylised version of a follicle and label it, it will help you remember these names.

primordial follicles

Earlier, very small stages like the primordial, primary and secondary follicles can be distinguished (Figure 2.12). The whole follicle population can be subdivided into resting and growing follicles. The resting ones are the primordial follicles, which form a pool. The number of follicles in the pool decreases very much just before or shortly after birth. This decrease is mainly caused by degeneration of primordial follicles but a small percentage shows growth.

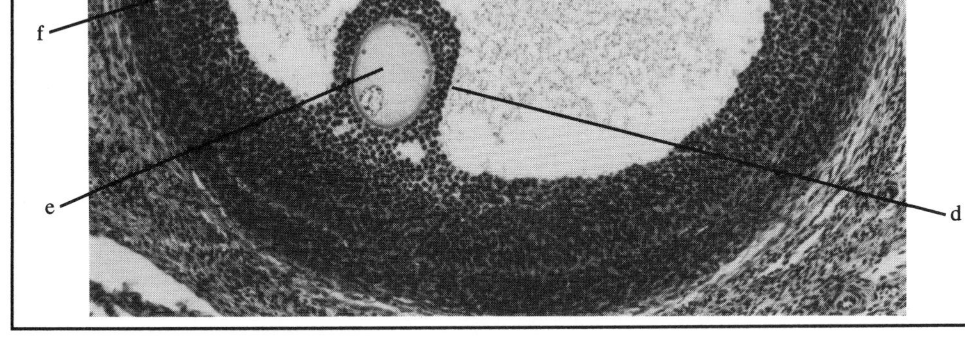

Figure 2.11 The micromorphological feature of a tertiary follicle. a) theca externa b) theca interna c) membrana granulosa d) cumulus oophorus e) oocyte f) antrum.

After the onset of puberty the decrease in numbers is only caused by a periodic growth and subsequent degeneration of follicles. The follicle grows until the moment degeneration starts or ovulation takes place.

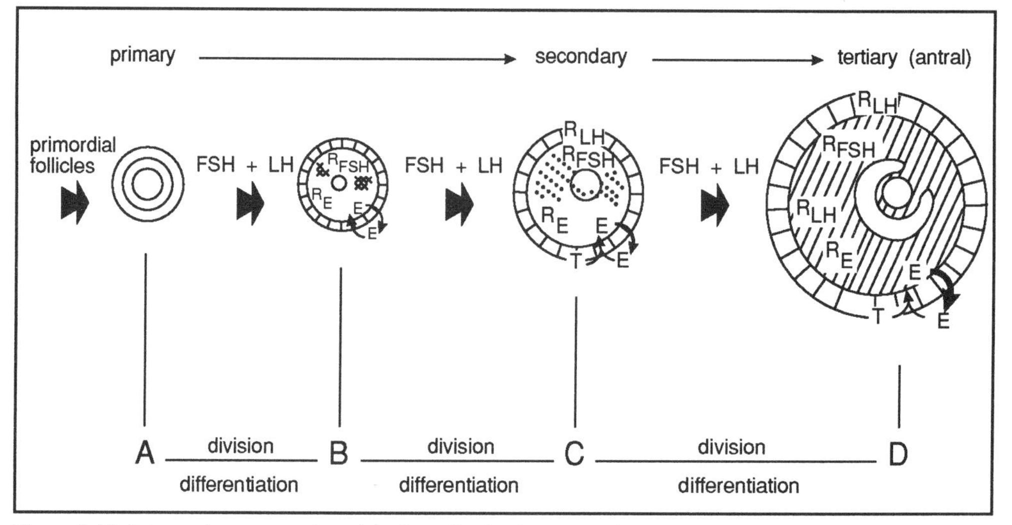

Figure 2.12 Schematic presentation of the hypothesis about growth and maturation of the ovarian follicles. RFSH, RLH and RE are symbols representing the receptors for FSH, LH and oestradiol-17β respectively.

The actual positions of the receptors are discussed in the text. Stages A-D are also described in the text. Note also that the tertiary follicle stage is also known as the antral stage. At this stage the follicle has a distinct antrum (see Figure 2.11).

The time between the beginning of follicular growth and its associated ovulation in sheep is nearly 6 months. Initially the growth is slow (phase A) but it increases rapidly thereafter (phase B). In phase A, it is the oocyte that grows, in phase B the follicle.

Π What stimulates the growth of the follicles? (Use Figure 2.12 to help you answer this).

FSH receptors

oestradiol-17β receptors

The initiation of follicle growth is independent of gonadotrophic hormones but the number of primordial follicles is directly proportional to the number of follicles that begin to grow. Once growth starts, the follicle needs the support of gonadotrophic hormones. During growth from primordial to primary follicle, receptors for FSH are formed on the membrane of the first layer of granulosa cells. Now the follicle is able to respond to the appearance of FSH in the blood. FSH is responsible for proliferation of the granulosa cells, for the formation of receptors for oestradiol-17β on the granulosa cells, and for the synthesis of a 'theca cell organiser' (secondary follicle). This hypothetical factor induces the formation of glandular cells in the connective tissue around the growing follicle. FSH is also responsible for the formation of follicular fluid and of the cavity which holds the fluid, the antrum (tertiary follicle).

LH receptors

androgens

oestrogens

Receptors for LH are formed on the glandular theca interna cells of the tertiary follicle. Under the influence of LH the theca cells synthesise androgens. The pathway of their synthesis is described in Figure 2.13. These androgens (mainly androstenedione) diffuse into the granulosa cells and become aromatised (ie aromatic rings are produced) to oestrogens (mainly oestrone and oestradiol-17β) under the influence of FSH. Oestrogens stimulate the proliferation of the granulosa cells, the production of the follicular fluid and the synthesis of inhibin. The granulosa cells differentiate into the mural layer of cells (the membrana granulosa) and the layers that surround the oocyte ie the cumulus oophorus with the corona radiata (Figure 2.11).

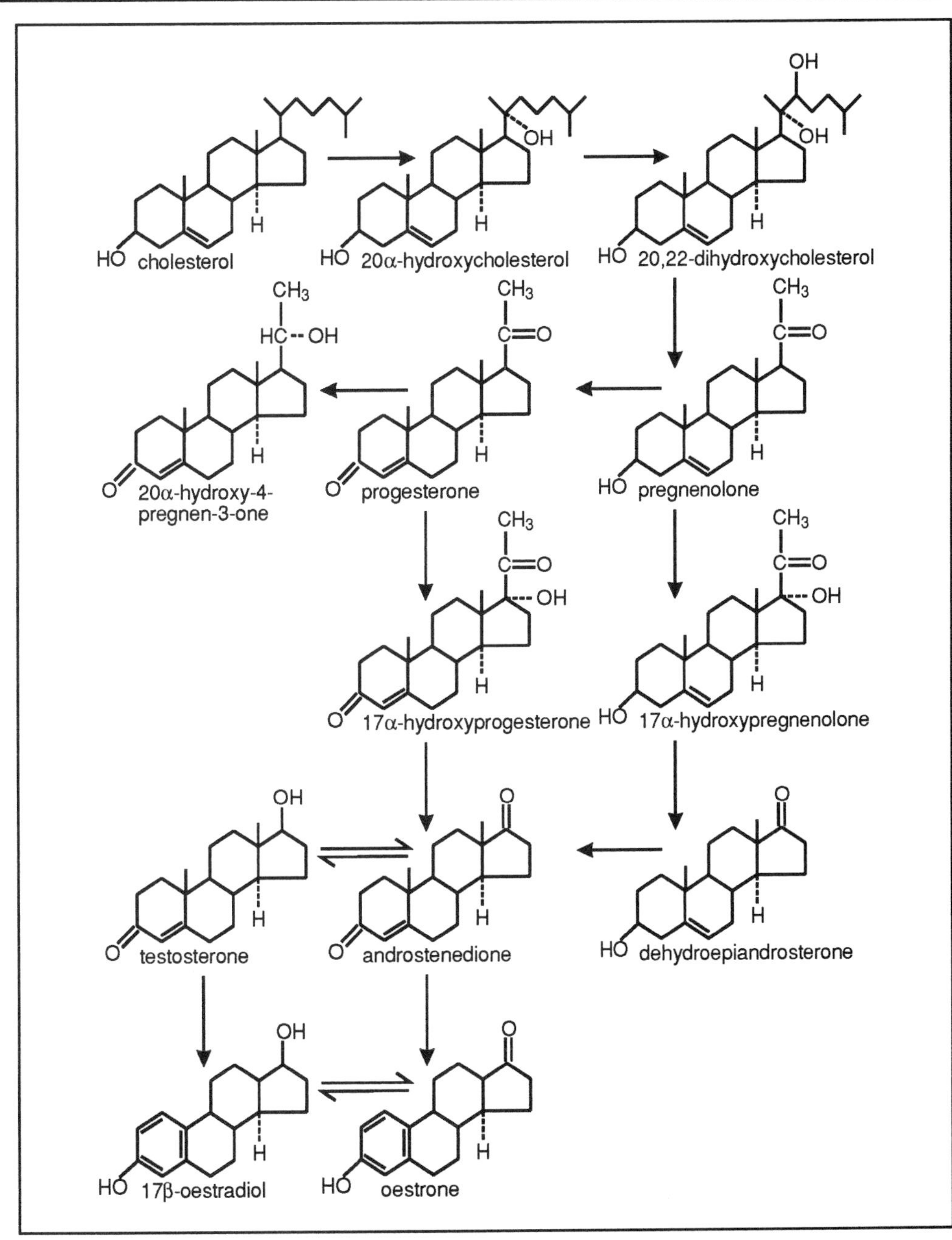

Figure 2.13 Steroid synthesis pathway in ovarian cells.

Π Identify in Figure 2.13 the main steroids so far discussed in the regulation of the oestrous cycle.

You should have picked out progesterone and oestradiol-17β. Mark on this figure which compounds diffuses from these cells to the granulosa cell. (You should have indicated androstenedione and possibly testosterone).

FSH and oestradiol-17β are responsible for the formation of receptors for LH on the granulosa cells. The moment that the oestradiol-17β synthesis reaches a maximum, the LH-peak occurs. Now the granulosa cells can respond to the LH. This leads to a diminution or a blockade of oestrogen synthesis. The granulosa cells 'luteinise' and start the production of progesterone. The progesterone is thought to be responsible for the synthesis of collagenase in the membrana granulosa. This enzyme facilitates the rupture of the follicular wall and release of the oocyte ie ovulation.

progesterone

ovulation

Π Again we would recommend you draw out a flow diagram of these hormonal events.

We have began drawing out a scheme, use the text to help you complete it.

primordial follicle ⟶ primary follicle ⟶ secondary follicle ⟶ tertiary follicle

FSH receptor develops;

FSH stimulate proliferation of granulosa cells;

FSH stimulate oestradiol-17β receptor development;

FSH stimulates the cell organiser.

Besides the production of steroid hormones, inhibin and many other factors, a very important function of the follicle is the creation of a favourable microenvironment for the oocyte.

In the primordial follicle the oocyte is in intimate contact with its surrounding granulosa cells. Development from a primordial stage into a primary follicular stage starts with the growth of the oocyte and the creation of space between oocyte and granulosa cells. A mucopolysaccharide substance is secreted between oocyte and granulosa cells.

zona pellucida

It is generally accepted that both cell types are responsible for the production of this substance which later forms the zona pellucida. The granulosa cells nevertheless keep contact with the oocyte by means of very thin cytoplasmic processes. This close contact between oocyte and surrounding cells breaks down in response to the LH-peak release. At the same moment the oocyte resumes meiosis. The meiosis of the oocyte has been initiated already in foetal life (we will discuss this in a later chapter), but halted in the diplotene of prophase I during the primordial follicular stage. Besides the resumption of meiosis as a first indicator of maturation, cytoplasmic maturation is induced by the LH-peak. At ovulation the cytoplasm is mature and the nucleus is in metaphase II stage. Now the oocyte must be fertilised within 4 to 6 hours. After fertilisation the meiosis will be completed with the extrusion of the second polar body and the male and female pronuclei will be formed.

atresia

Most oocytes do not leave the ovary by ovulation but by degeneration of the follicles. Follicles can start degeneration at any time during growth but mostly do so in the tertiary follicular stage. Very little is known about the real cause of this degeneration or atresia, but a shortage of LH and FSH is involved. Administration of additional LH and FSH by injection leads to superovulation (multiple ovulations). Blockade of LH and FSH release by injection of Nembutal (barbiturate) leads to an increase of the number of atretic follicles. However these observations do not explain why one follicle survives and another degenerates. The cause may be in the follicle itself. Some follicles possess more receptors for gonadotrophic hormones than others.

It is clear that the ovarian follicles play a very important role in the regulation of the oestrous cycle by: a) the synthesis of oestradiol-17β, and shortly, before ovulation, by progesterone synthesis; b) the synthesis of factors inhibiting and stimulating oocyte maturation; and last but not least, c) the synthesis of factors that are involved in ovulation and formation of the corpus luteum. All these processes are dependent on critical concentrations of gonadotrophic hormones being present in the blood.

After ovulation the luteinisation of the follicle proceeds to form the corpus luteum. In domestic animals the principal steroid produced by this organ is progesterone.

As already stated, progesterone has a negative feedback effect on the tonic release of LH (Figure 2.6). This means that growth of the corpus luteum is followed by decrease in LH concentration in the blood until a balance is reached (Figure 2.4). Follicles growing during the early luteal phase can also produce reasonable amounts of oestrogens but oestradiol-17β production by follicles during the mid-luteal phase is very low (Figure 2.4). If the corpus luteum regresses and the progesterone concentration decreases, the frequency of LH-pulse release becomes enhanced and the oestradiol-17β production increases to maximal values. This results in the induction of oestrous behaviour, including LH-peak release and a new ovulation.

2.6.2 Lifespan of the corpus luteum

After ovulation the luteinisation of the follicular wall proceeds and forms the luteal tissue or corpus luteum. When the oocyte becomes fertilised, the embryo develops and the animal becomes pregnant, the corpus luteum will continue to function for a long, but species specific, period. If the animal is not mated or does not become pregnant the corpus luteum soon regresses, allowing another oestrus and ovulation, and a second chance of fertilisation.

prostaglandin-F2α

Prostaglandin-F2α (PGF2α) is a locally acting mediator with hormone-like activity involved in the regression of the corpus luteum. The endometrial cells of the uterus synthesise and secrete PGF2α when no embryo arrives to implant in the uterus. As will be described in Chapter 3, prostaglandin is a potent drug that kills (luteolyses) the corpus luteum if injected. It acts as a vasoconstrictor and can bind to the LH-receptors. In the young corpus luteum the LH-receptors are fully saturated with LH. Hence in the early luteal phase of the cycle prostaglandin injection has no effect. In the mature and older corpus luteum, not all LH-receptors are occupied. Prostaglandin-F2α binds to them preventing the LH stimulation of adenylate cyclase and cAMP production, and as a consequence the production of progesterone. The concentration of the progesterone in the blood decreases and a new oestrous cycle can start.

2.7 The genital tract

The genital tract of large domestic animals, consisting of paired fallopian tubes and uterine horns, and a single cervix and vagina, is one of the main target organs of the ovarian steroid hormones.

Cyclical variations are evident both morphologically and biochemically in all parts of the genital tract. They can easily be related to the cyclical variations in the concentrations of oestradiol-17β and progesterone in the blood. They contribute to the changing microenvironment of the tract to ensure the optimal transport of gametes, fertilisation and development of embryo and foetus.

If no embryo arrives in the genital tract, the changes are also linked to the production and release of prostaglandin-F2α, the 'hormone' that indirectly induces the start of a new oestrous cycle.

SAQ 2.5

Select the right functions from the list provided for each of the following hormones.

	Hormone	Functions
1)	Luteinising hormone (LH)	stimulate lactation
		oocyte maturation
2)	Follicle stimulating hormone	support corpus luteum
		stimulate steroid synthesis
3)	Prolactin	ovulation

SAQ 2.6

Indicate which of the following are true and which are false.

1) FSH is responsible for the proliferation of granulosa cells during follicle development.

2) FSH is responsible for the formation of receptors for oestradiol-17β on the granulosa cells.

3) LH receptors are formed on the glandular theca interna cells during follicle development.

4) LH stimulates the synthesis of androgen by the theca cells.

5) Androgens diffuse from the theca cells to the granulosa cells in the follicle where they are converted to oestrogens.

6) Mucopolysaccharide is secreted between oocyte and granulosa cells as the follicle develops from the primordial to the primary state.

7) LH-peak release is associated with the resumption of meiosis by the oocyte.

8) LH-peak release is associated by the severing of the cytoplasmic bridge between the oocyte and the surrounding cells.

9) The degeneration of follicles is known as atresia.

10) Follicle production of oestrogens is low during the mid-luteal phase.

11) Prostaglandin-F2α causes the destruction of the corpus luteum only if the corpus luteum has LH-receptors which are unoccupied by LH.

2.8 Summary

Taking the sheep as our example, many aspects of the regulation of reproduction can now be explained.

The anoestrous season is the consequence of an increasing sensitivity of the hypothalamic nuclei to the negative feedback of oestradiol-17β under the influence of increasing day-length (environmental cue). Due to this negative feedback mechanism the release of LH by the pituitary gland decreases progressively. Decline in LH release leads consequently to reduced oestrogen production. At latitude around 51°, the synthesis of oestradiol-17β falls to a level insufficient to induce a LH surge (peak release). Hence no oestrous behaviour or ovulation can be detected from mid-February onwards (anoestrous season).

Use Figure 2.14 to help you follow the description given below. Note that only three hormones levels are shown (progesterone, luteinising hormone and oestradiol).

When day-length starts to shorten from mid-late June onwards, the sensitivity of the hypothalamic nuclei to the negative feedback of oestradiol-17β decreases and consequently more LH is released. More follicles start growing and reach the preovulatory size with maximal oestradiol-17β synthesis (Figure 2.14). In early September the first endogenous LH-peak release occurs and ovulation follows. Since during the anoestrous season neither ovulations nor formation of corpora lutea occurred, the ewe did not receive any pretreatment with progesterone and does not respond to the oestradiol-17β concentration in the blood with oestrous behaviour. The oestrous season therefore starts in September with a silent heat. Although ovulation takes place and a corpus luteum will be formed the female is not mated, so no embryo develops in the uterus and prostaglandin-F2α is released by the endometrium. This sequence of events is followed by regression of the corpus luteum, and a decrease of progesterone in the blood. An increase of the tonic LH-pulse release occurs and a new ovulation is induced. When the ewe is served by a ram, the oocyte is fertilised and the embryo develops in a pregnancy lasting for nearly 5 months. During pregnancy the high progesterone concentration in the blood blocks oestrus and sexual activity. If the ewe does not become pregnant during the autumn the oestrus cycle continues until increasing day-length raises the sensitivity of the hypothalamus to the negative feedback of oestradiol-17β. No further LH-peak releases occur, ovulation ceases and the ewe enters its anoestrous season (Figure 2.14).

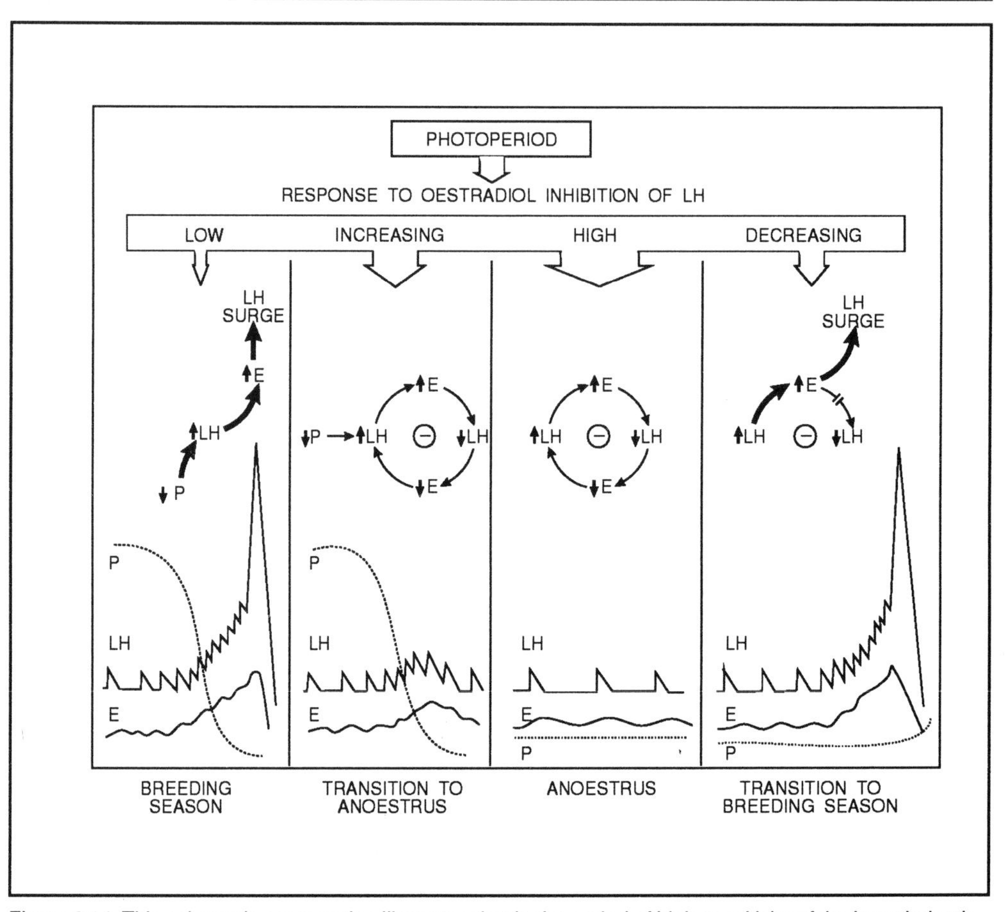

Figure 2.14 This schematic presentation illustrates that in the period of high sensitivity of the hypothalamic nuclei to the negative feedback of oestradiol (anoestrous season) the LH pulse release has a low frequency. As soon as the day-length shortens and the sensitivity of hypothalamic nuclei to the negative feedback decreases, more LH is released, more oestradiol is synthesised and oestrous behaviour and LH-peak release are induced.

SAQ 2.7

What is a silent heat and what function might it serve in the sheep?

Summary and objectives

This chapter explained the basic information available on the oestrous cycle of mammals, particularly for domestic species like the sheep. It described and explained the cyclical nature of oestrus, and of breeding seasons on the basis of the influence of environmental factors such as photoperiod, and internal regulatory mechanisms.

Now that you have completed this chapter you should be able to:

- recall key terms relating to the regulation of reproductive cycles;
- understand the nature of the oestrous cycle in large domestic animals;
- assign roles to particular hormones in the regulation of the oestrous cycle;
- appreciate the roles of environmental influences and their seasonality, the central nervous system, endocrine glands and feedback processes in regulating reproduction.

Manipulation of reproduction

Manipulation of reproduction

3.1 Introduction

production targets with cattle

production targets with sheep

Profitability of animal production depends to a large extent on reproductive performance. For cattle this means the production of a calf per cow each year. If the breeding cow does not show regular cyclic breeding activity, become pregnant at the appropriate time and deliver a live, healthy calf each year, then her other excellent qualities may be to no great avail. For sheep the production of many live lambs is the optimum. Reproductive efficiency is usually measured as the lambing percentage, the number of lambs born alive or surviving to weaning, per 100 ewes mated. Differences between breeds and type of husbandry system are seen. Whereas lambing percentages of around 80%-85% are considered acceptable in extensive range systems, under intensive confinement conditions, figures of 170%-200% can be achieved. Productivity may be increased by out-of-season breeding to produce two lamb crops a year, and/or by increasing the proportion of twins born.

production targets with pigs

For pigs, which can produce several litters a year, maximum productivity is achieved if the gilts reach puberty before eight months of age and sows return to oestrus within ten days of weaning. Typical production targets might include: a farrowing rate (number of sows farrowing per 100 services given to the herd) of 89%; 11 piglets born alive per litter; and production of 23 pigs per sow per year. So, the importance of reproduction as a factor influencing the efficiency of animal production on the farm should require no further emphasis.

In the years ahead, the application of science (including biotechnology) to animal agriculture will become increasingly important in meeting world-wide requirements for animal products. Controlled breeding, in one form or other, can be expected to contribute substantially in improving the efficiency of animal production. In terms of improving animal quality, for example, artificial insemination has already had a major impact on dairy cattle breeding in many countries. We should keep in mind that although we may have production targets, the 'quality of life' of domestic animals is of great concern to many and is an important consideration in animal production. We will not however examine this aspect further here.

This chapter therefore focuses onto the techniques which are used to manipulate reproductive processes in the female animal. They are the result of expanding knowledge of reproductive physiology and biotechnology.

3.2 Pharmacological control of oestrus and ovulation

Increased understanding of the physiological mechanisms controlling reproduction has led to procedures which may now be employed in many commercial situations to improve the existing reproductive performance of cattle, sheep, pigs and horses.

Π Write down a list of the areas you think we need to understand in order to be successful in controlling the oestrous cycle pharmacologically?

We anticipate that you probably come to the following conclusions.

The success of oestrus control programmes depends on an understanding of three general areas:

- physiology of the oestrous cycle;
- pharmacological agents and their effect on the oestrous cycle;
- herd management factors that may reduce anoestrus and increase conception rates.

The complex nature of the hormonal control of the female reproductive system, described in the previous chapter, means that there can be several different possible causes for a particular problem. For example, undetected oestrous could be due to a hormonal defect at the level of the hypothalamus, the pituitary or the ovary.

artificial insemination or AI

In addition, man-made management systems have added another layer of complexity. Use of artificial insemination (AI) in cattle, for example, means that Man rather than the bull must be able to detect signs of oestrus. Thus, an animal may have normal ovarian cycles, but not be detected to be in heat by farm staff. The failure of such an animal to breed is therefore a management rather than a physiological problem.

The intensification of farming in the US and Europe, with animals being kept in larger groups, has added to the difficulties of management. Farm staff may have less time for oestrous observation, and animals may be less likely to express it if conditions are not favourable.

Pharmacological agents are very useful in oestrus synchronisation programmes but they cannot replace good management. So, close attention to management and good nutrition are necessary for successful pharmacological control of oestrus and ovulation.

3.2.1 Basic principles of hormonal manipulation

The events of the oestrous cycle are coordinated for the following major functions:

- development and ovulation of a mature, viable oocyte;
- induction of oestrous behaviour during the critical period when fertilisation can occur;
- preparation of the oviduct and uterine environments for their roles in sperm transport and capacitation, oocyte and embryo transit from ovary to uterus, and the successful establishment of pregnancy;
- reinitiation of the entire sequence should pregnancy fail at any stage.

hypothalamo-pituitary-ovarian axis

Control of the oestrous cycle involves the interrelated secretion of a number of hormones from the hypothalamo-pituitary-ovarian axis.

Π Make a list of the main hormones involved?

From your reading of Chapter 2 you should have been able to write quite a list of hormones. They include GnRH from the hypothalamus; FSH and LH from the pituitary gland; oestrogen, progesterone, and inhibin from the ovary; and PGF_{α} from the uterus. If you could not remember most of these, perhaps you need to do a little more work on Chapter 2. The involvement of additional hormones cannot be excluded as we do not know all of the details of hormonal regulation.

A number of pharmacological treatments, designed to mimic or replace the natural hormones involved in reproduction, are available. Their application to the control of reproduction is discussed below, and a summary of the various agents is given in Table 3.1.

Natural hormone	Source	Pharmacological product
Gonadotrophin-releasing hormones (GnRH)	Hypothalamus	- chemically synthesised GnRH - GnRH-analogues, which are slightly different chemical structures
Luteinising hormones (LH)	Pituitary	- human chorionic gonadotrophin (hCG)
Follicle-stimulating hormone (FSH)	Pituitary	- equine chorionic gonadotrophin (eCG), also called: pregnant mare's serum gonadotrophin (PMSG) - FSH from pituitary extracts
Oestrogen	Ovary	- synthetic oestrogens
Progesterone	Ovary	- synthetic progesterone - progestagens, ie structural analogues of progesterone
Prostaglandin-F2α	Uterus	- synthetic prostaglandin- F2α - analogues of prostaglandin-F2α

Table 3.1 Natural hormones involved in the control of female mammalian reproduction and pharmacological agents which may replace their action.

Π Examine Table 3.1 closely and make certain you can remember what hCG, eCG and PMSG stand for as we will be discussing these later.

3.2.2 Synchronisation of oestrus and ovulation

In the cyclic animal, control of the timing of oestrus and ovulation is dependent on controlling the time of regression of the corpus luteum. Therefore, the two main methods of artificially controlling the time of oestrus and ovulation are:

- the induction of premature but predictable regression of the cyclic corpus luteum with PGF2α;
- the holding of animals in an artificial luteal phase with exogenously administered progesterone until endogenous regression of the corpus luteum has occurred in all animals.

These two types of approach are outlined in Figure 3.1. Of course, combinations of the above are also possible used in conjunction with various management techniques.

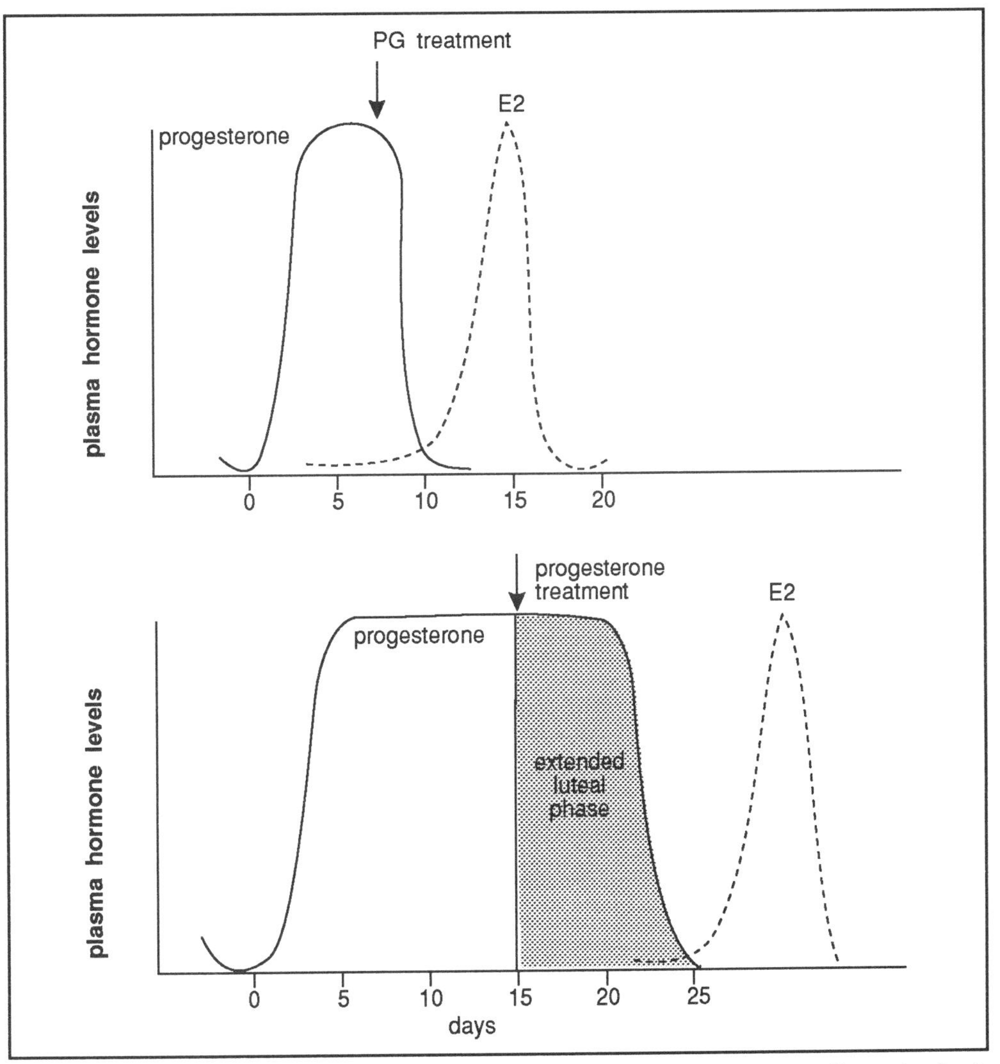

Figure 3.1 Synchronisation of oestrus with prostaglandin (PG) or progesterone treatment. Plasma levels of progesterone represent the luteal phase and oestradiol (E2) levels represent the follicular phase.

Π With the first technique, do you think that animals will always respond to a prostaglandin treatment?

Prostaglandins, either PGF2α or its synthetic analogues, will cause regression of a functional corpus luteum providing their LH-receptors are not blocked by LH. This means prostaglandins are only effective between days 5 and 17 of the cow's cycle. Following a single prostaglandin injection, cows which respond will return to oestrus within five days (normally 60-72 hours) of treatment and can be inseminated at the observed oestrus. Thus, only if the stage of the cycle is known, can prostaglandin treatment by a single injection be applied.

Since prostaglandins are only effective when a functional corpus luteum is present two serial injections are required in order to synchronise a group of cows which are cycling at random. The first prostaglandin injection will not be effective in cows in the early luteal phase (before day 5 of the cycle), nor in those animals which are already in the follicular phase (the last 3 to 4 days of the cycle). This will be approximately 45% of the animals in a randomly cycling group.

syncrhonised oestrus

However, 10-12 days after the first injection 90 to 95% of animals should have a functional corpus luteum, and so be sensitive to prostaglandin. A second injection given at this stage thus induces the onset of synchronised oestrus, which occurs two or three days after the injection.

Fixed time AI

Fixed time AI without oestrus observation, can be given following prostaglandin synchronised oestrus. The recommended procedures are either a single insemination, 60 to 72 hours after the second prostaglandin injection, or two inseminations, at 72 and 96 hours.

By contrast, in pigs prostaglandin treatment is of little use for oestrous control, as the corpus luteum is only responsive to treatment between days 12 and 15 of the twenty-one day oestrous cycle.

progestagens

In the second technique for synchronising oestrus and ovulation, progestagens (progesterone-like substances) can be used to mimic corpus luteum function. Negative feedback by the progestagen will inhibit the release of gonadotrophins by the pituitary gland. However, the gonadotrophins will be released once the progesterone blockade is removed, and this may result in oestrus and ovulation.

progestagen delivery systems

Some presently available delivery systems for progestagens are:

- a sponge, impregnated with a progestagen. While the sponge is in place the progestagen is absorbed through the vaginal wall (see Figure 3.2);
- a silicone rubber implant containing the progestagen norgestomet, and which is placed subcutaneously in the ear of cattle;
- a stainless steel spiral device covered with silicone rubber which contains progesterone. It is used as an intravaginal device in cattle;
- the progestagen allyltrenbolone is given orally in the feed and used to control the ovarian cycle in mares. It may also be used for oestrus synchronisation in pigs. The initial approaches of administering progesterone or progestagens for 18-20 days resulted in good control of the timing of oestrus but lower calving rates than in untreated animals. This low fertility has since been shown to be attributable to the duration of treatment with the progestagen. When treatments of 9 to 12 days were used, fertility was essentially normal.

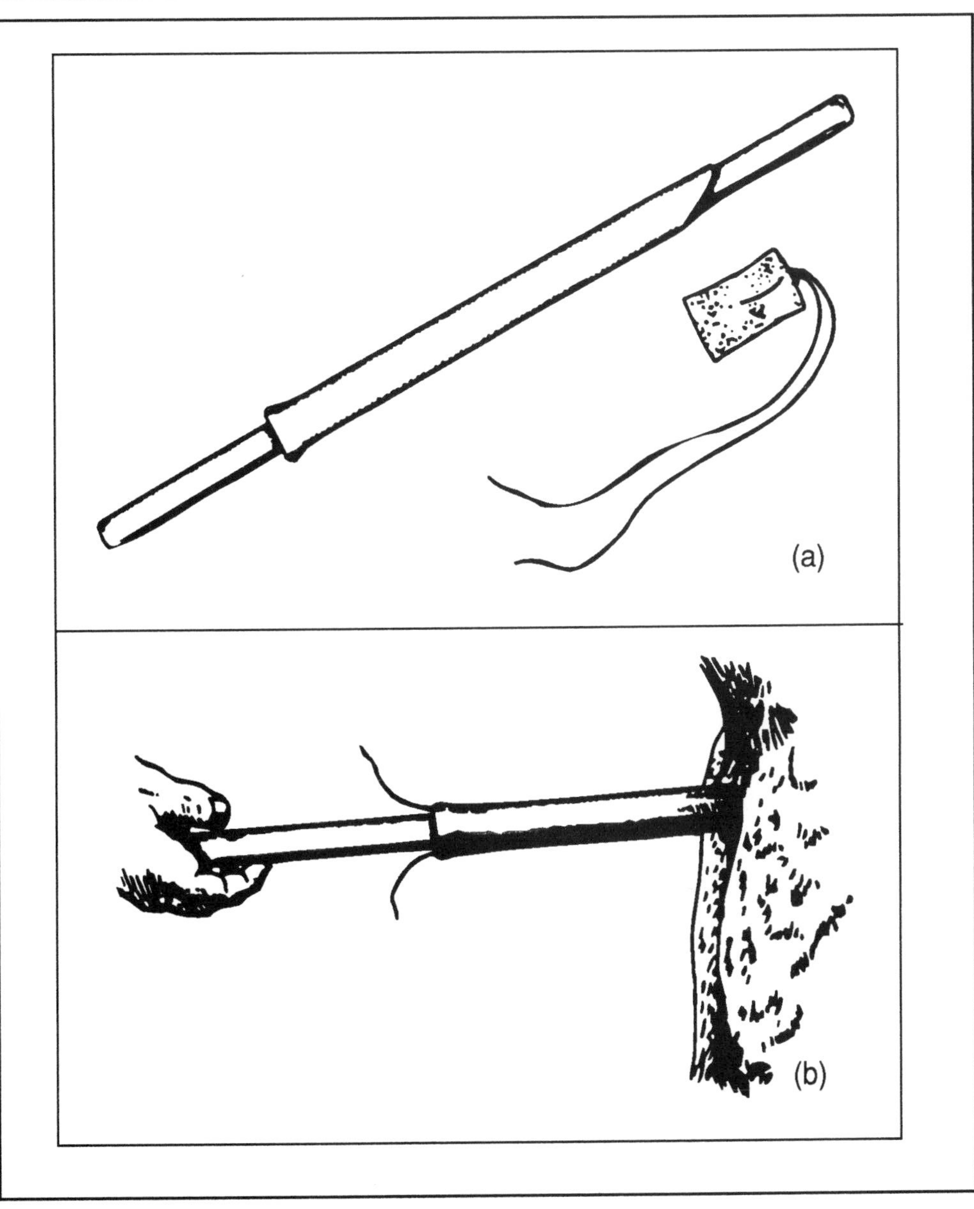

Figure 3.2 Intra-vaginal insertion of a sponge in sheep and goats.

Π Is a synchronised oestrus likely to be obtained when cattle are treated only with progestagens for 9 to 12 days? (Examine Figure 3.1 and remember that to begin with all the animals in a herd will be at different stages in the oestrous cycle. Also remember that the oestrous cycle is 21 days long in cattle).

The major part of the 21 day-long oestrous cycle is dominated by the presence of progesterone secreted by the corpus luteum. So, when progestagen treatments of 9 to 12 days are used, the life-span of the existing corpus luteum has to be shortened in order to achieve a high level of oestrus synchronisation. Therefore, the presently applied progestagen treatments in cattle include the administration of a luteolytic agent. Since oestradiol induces uterine luteolytic mechanisms so as to cause regression of the existing corpus luteum, this steroid hormone is administered at the onset of the progestagen treatment. On the other hand luteolysis might also be induced by injection of prostaglandin 1-2 days before the end of the progestagen treatment.

3.2.3 Induction of oestrus and ovulation

The most practical method to induce oestrus and ovulation in cattle (eg post-partum cows which are in anoestrus or suboestrus) is to use a progesterone or progestagen treatment of 9-12 days duration and to give an injection of exogenous gonadotrophin (PMSG = pregnant mares serum gonadotrophin) at the end of the progestagen treatment. This treatment will induce ovulation in most anoestrous cows. The dose of PMSG used varies with breed, season and nutritional status of the cow.

3.2.4 Advancement of the breeding season in sheep and goats

melatonin

Unlike cattle, sheep and goats have a breeding season which usually begins in late summer and continues through until the start of spring. From research with light manipulations it became clear that there is a relationship between decreasing photoperiod, ie short days, and increased reproductive activity in sheep. It also appeared that these short days were accompanied by increased secretion of the hormone melatonin from the pineal gland. Moreover, administration of melatonin for several weeks appeared to mimic the effects of shortening day length on reproductive activity. Recently, an implant has been developed which releases the pineal hormone melatonin for 6 weeks. The implant, which is placed subcutaneously in the neck, can be used in sheep to advance the breeding season. However, additional research is required before the melatonin treatment (subcutaneously as well as orally) can be optimally applied as an aid in the reproductive management of sheep.

Control of oestrus and ovulation in sheep and goats, whether in the full breeding season or in the anoestrus season, is usually based on attempts to simulate the activity of the corpus luteum. Therefore, in both species, sponges impregnated with a progestagen can be used to synchronise oestrus or to advance the breeding season. During the breeding season it is not necessary to give the progestagen treatment with gonadotrophin injection. However, in the presence of the gonadotrophin, a more predictable and precise synchronisation of oestrus and ovulation is obtained. On the other hand, for out-of-season breeding the progestagen treatment must be combined with the administration of PMSG around the time of sponge removal in order to induce ovulation. The dose of the exogenous gonadotrophin used should be sufficient to induce ovulation. Dosage of PMSG required depends on season and breed.

Under normal circumstances we do not want to use too much PMSG as this will cause multiple ovulations. This in turn may lead to multiple pregnancy, low lamb weight and thus poor survival rates. There are however some circumstances in which superovulation is desirable. We will deal with these in the next section.

SAQ 3.1

What common basis links the two main techniques described for synchronisation of oestrus and ovulation?

3.3 Superovulation and embryo transfer

Making greater use of the egg cells in the ovaries of the genetically superior animal remains one of the main objectives of research in animal reproduction. Although the dream of maturing at will the many thousand of egg cells in the ovaries has not yet been realised, there are interesting developments in this area that permit greater use of the female. One successful approach involves the hormonal induction of superovulation (multiple ovulations) in an embryo transfer programme. This method has moved from the laboratory to the farm and is widely used in cattle. The advantages of embryo transfer include:

superovulation

embryo transfer

- production of multiple offspring from genetically superior parents;
- increasing selection pressure in bull testing for use in artificial insemination programmes;
- production of disease-free animals for import/export.

In addition, embryos are obviously much cheaper to transport than live cattle. When embryos are transferred to native recipients the transplanted calf will acquire immunity to local diseases from its foster-mother.

In cattle, two approaches have been used to obtain a supply of eggs. The first involves the hormonal induction of superovulation and the second approach entails the recovery, maturation and fertilisation of eggs taken directly from the ovaries. The first method is discussed below and the second approach will be reviewed in the next chapter.

3.3.1 Superovulation

gonadotrophin treatment

The first step in cattle superovulation treatments is the administration of an appropriate FSH-type gonadotrophin two days in advance of oestrus (natural or predetermined oestrus). The two most important gonadotrophins currently used are pituitary extracts of FSH and PMSG (eCG) (see Chapter 2 and Table 3.1). The FSH preparations are regarded as having a short period of activity and for that reason have usually been administered over a period of 4 or 5 days with 2 doses per day (ie a total of 8 or 10 injections). On the other hand, PMSG appears to have an unusually long biological half-life (estimated at 5-7 days in cattle), so only one injection is needed.

The cow seems to superovulate best if gonadotrophin treatment is begun around the mid-luteal stage of the cycle. As shown in Figure 3.3 a method of superovulating cattle involves the administration of the gonadotrophin PMSG in a single dose during days 8-13 of the oestrous cycle, followed 48-72 hours later by a luteolytic dose of prostaglandin. Normally the donor cow can be expected to come into oestrus 2 days later and then can be inseminated.

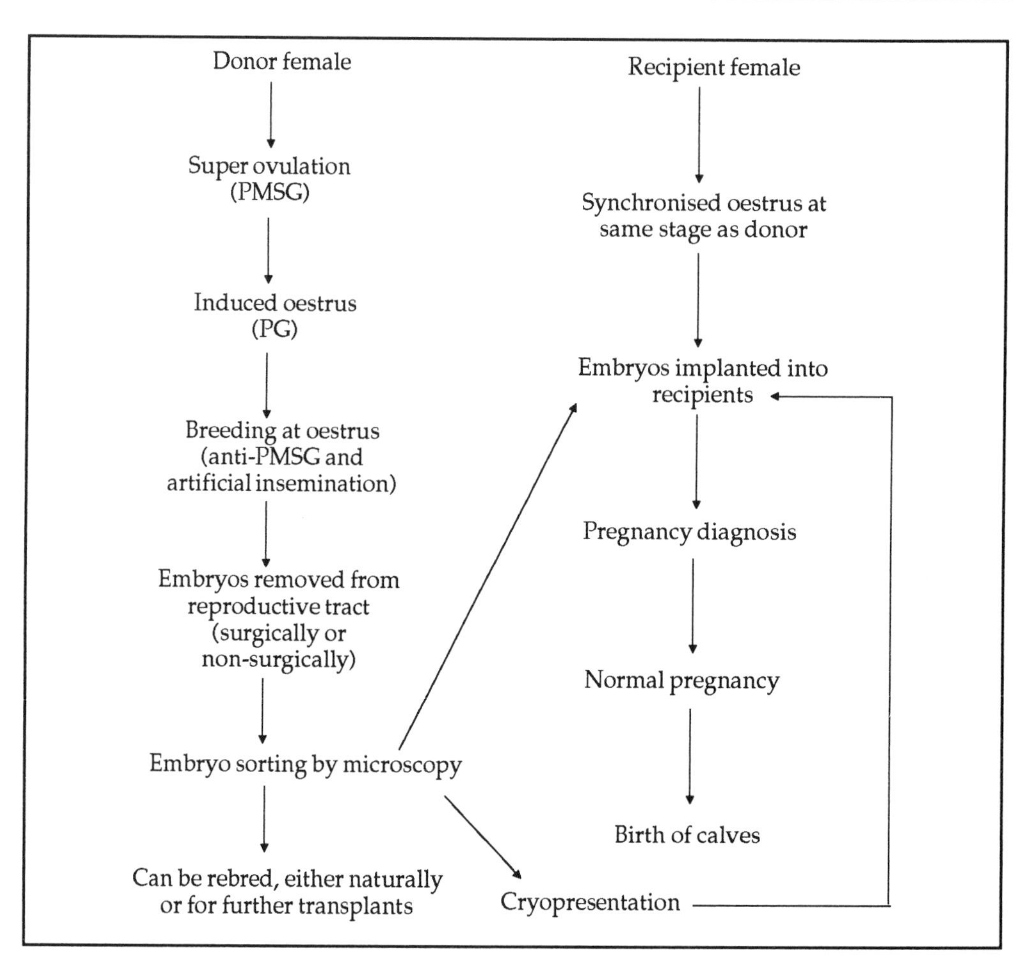

Figure 3.3 Embryo transplantation in the cow. between 60% and 90% of cows respond by superovulating, and the average number of embryos collected from each donor cow is four to five. The response to superovulation is very variable however, and the numbers of embryos obtained per cow can range from none to 20. (see text for details).

antibodies

anti-PMSG

As a result of the many studies involving PMSG, it was felt by some researchers that the presence of PMSG in the circulation of the cow after the time of multiple ovulation might have an adverse effect, particularly on the quality of the developing fertilised eggs. In an effort to interrupt the otherwise prolonged action of PMSG animals were treated with a serum containing antibodies against PMSG around the time of insemination. From these studies it appeared that the use of anti-PMSG improved the development of the embryos, both qualitatively and quantitatively. See also Section 3.6 of this chapter.

3.3.2 Recovery of embryos

Embryos of domestic animals do not implant into the uterine tissue until some time after fertilisation. It is this fact that makes embryo recovery and transfer possible. In cows embryos can be recovered by either surgical or non-surgical techniques. Non-surgical collection is the simpler procedure, and can be conducted on the farm. Embryos are flushed from the reproductive tract using a catheter introduced via the cervix. Best results are achieved with 6-8 day embryos. The embryos are collected and examined under a microscope in order to check that they are morphologically normal.

3.3.3 Transfer of embryos

Embryos can either be transferred directly to a recipient animal, or be frozen and kept in liquid nitrogen for later transfer. Embryos can also be manipulated at this stage, for example by splitting to produce identical twins (see Chapter 4).

Π It might be fun to redraw Figure 3.3, but to make a pictorial version of it. Begin with a picture showing a cow being injected with PMSG. This is an excellent way of remembering the sequence.

Both superovulation and oestrus synchronisation are critical components of an embryo transfer programme. Recipients can be selected either by detection of natural oestrus in untreated animals or by detection after drug-induced oestrus synchronisation as described in Section 3.2.2 of this chapter. Regardless of the method of synchronisation used, timing and careful attention to oestrus detection are important.

3.3.4 Embryo transfer in other farm animals

Embryo transfer is also possible in sheep, goats, pigs and horses, but is not used to anything like the same extent as in cattle. Surgical transfer has to be used in pigs and is also usual in sheep and goats.

SAQ 3.2

Should it be possible to implant an embryo into the recipient at any stage of the oestrous cycle?

3.4 Management of pregnancy and parturition

3.4.1 Pregnancy

progesterone

Progesterone is required throughout gestation in all mammals in order to maintain the pregnancy. There are however, differences between the species in the source of the hormone. In the cow, goat and pig the corpus luteum persists throughout gestation, and is the major source of this hormone. In the sheep and horse, on the other hand, the placenta takes over production of progesterone after approximately one-third of the gestation period.

Since the corpus luteum plays an important role during pregnancy, it should be apparent that pregnancy will terminate when the function of the corpus luteum is abolished. In some situations it may be necessary to terminate pregnancy in farm animals. On the other hand, when the corpus luteum does not secrete sufficient progesterone, pregnancy failures will occur. Administration of exogenous progesterone during pregnancy is therefore recommended.

Π Suggest a method that uses a hormone-like agent to terminate pregnancy in cattle.

From your knowledge of techniques used to synchronise oestrus (Section 3.2.2) it should be clear that prostaglandin treatment would be effective. Injection of PGF2α or its analogues from day 5 to day 100 terminates pregnancy in cattle by causing corpus luteal breakdown.

3.4.2 Parturition

PGF2α

Pharmacological agents may be used to induce (synchronise time of) parturition in farm animals, for example in pigs. The pig is dependent on the corpus luteum throughout gestation. Farrowing may be induced in sows by injection of prostaglandin, (either PGF2α or a synthetic analogue) and this method is now widely used in commercial pig farms. Most treated sows will farrow within 12 to 32 hours of treatment. The use of prostaglandin for this purpose requires good management and record keeping, as the injection should be given no more than 72 hours before the expected time of natural parturition. This is in order to decrease the chances of stillbirths. Oxytocin may be given following prostaglandin treatment in sows to give greater synchronicity in farrowing.

3.5 Immunological methods to manipulate reproduction

active and passive immunisation

The complex inter-relationships between the hormones involved in female reproduction can be altered in a number of ways as shown in previous sections of this chapter. One novel approach is to use antibodies to one or more of the components of the control system. Antibodies, by binding to reproductive hormone(s), will alter the positive and negative feed back loops. Antibodies, which do bind to a hormone and as a consequence neutralise its action, may be obtained either by active immunisation or by passive immunisation.

Π What are the differences between active and passive immunisation?

Active immunisation means that the animal is injected with the antigen, and as part of the response of the immune system antibodies are produced by cells called lymphocytes. Passive immunisation means that exogenous antibodies to a particular antigen are administered to the animals. These antibodies have been produced by laboratory animals or in *in vitro* systems as will be described below.

Usually the immune system will not respond to hormones, which occur naturally in the animal. So, the hormone has to be made immunogenic. This may be achieved by coupling the hormone to an immunogenic carrier protein. Moreover, to trigger the immune system adequately an adjuvant (see Chapter 6) is required.

3.5.1 Stimulation of reproductive activities

Fecundin

The first commercial product based on the use of immunological methods to manipulate the hormonal control of reproduction is Fecundin. This increases the number of eggs produced and thus increases the proportion of multiple pregnancies in sheep.

Fecundin is a conjugate of the ovarian steroid androstenedione with the protein human serum albumin (ie androstenedione linked to human serum albumin), with DEAE-dextran as adjuvant. Although its mechanism of action is not fully understood, it might be explained as follows.

Following injection with the product, ewes develop antibodies to the androstenedione. These antibodies bind some of the ovarian steroids involved in the control of reproduction. This reduces the feedback of ovarian steroids on the hypothalamus and pituitary, and results in increased production of gonadotrophins. In about 40% of ewes two follicles develop, rather than one, resulting in twin ovulations. We should not, however, exclude a direct effect as immunisation against ovarian steroids may cause a shift in the intraovarian hormonal control of the number of ovulations.

From recent research it has become clear that an increase in ovulation rates could also be obtained by immunisation (both actively and passively) against other ovarian steroids. From these studies and from the work with Fecundin, it also appears that the antibody levels are critical, ie too high levels leads to adverse effects, while too low levels do not affect the reproductive system at all.

3.5.2 Suppression of reproductive activities

immuno-castration

Suppression of reproductive activity in heifers and cows may be accomplished by administration of progestagens for a long period, eg oral administration in feed. However, reproductive activity can also be suppressed by immunisation (either passively or actively) against GnRH or LH. This technique, called immunocastration, can be applied in both males and females, and may be an alternative to surgical castration.

Immunocastration vaccines or antibodies to neutralise the actions of GnRH or LH are not yet commercially available.

3.6 Monoclonal antibodies as an aid to control reproduction

hybridoma technology

monoclonal antibody

In 1975 Köhler and Milstein developed the so-called hybridoma technology. Their starting point was the knowledge that a single lymphocyte (B-cell) produces a single antibody and that normal lymphocytes only manage to divide a few times before they die, but that cancer cells are immortal. If a lymphocyte, taken from the spleen where such cells are produced, could be fused with a cancer cell and then cultured, the result would be a clone of hybridised cells, ie a large number of identical cells called hybridomas. These would all secrete a single identical (or monoclonal) antibody. Such cells could be grown in large-scale cultures to provide the specific antibodies. (Note a detailed description of monoclonal antibody production is provided in the BIOTOL text 'Technological Applications of Immunochemicals').

Possible applications of monoclonal antibodies in livestock husbandry are:

- diagnostic test kits;
- prevention and treatment of viral and bacterial infections;
- elimination of drugs and toxins;
- manipulation of physiological processes, such as regulation of growth and reproduction (as mentioned briefly in Section 3.5.1).

An example of the use of monoclonal antibodies in test kits is discussed below. As we described in Section 3.3.1, monoclonal antibodies (anti-PMSG) can also have applications in superovulation. Monoclonal antibodies against PMSG neutralise residual circulating PMSG during a superovulation treatment.

3.6.1 Production of monoclonal antibodies

Until recently the method of choice to produce monoclonal antibodies was the production of ascitic fluid in mice. Hybridoma cells were injected intraperitioneally, and ascitic fluid was collected after two to three weeks. In affect, the animals were used as 'culture chambers'.

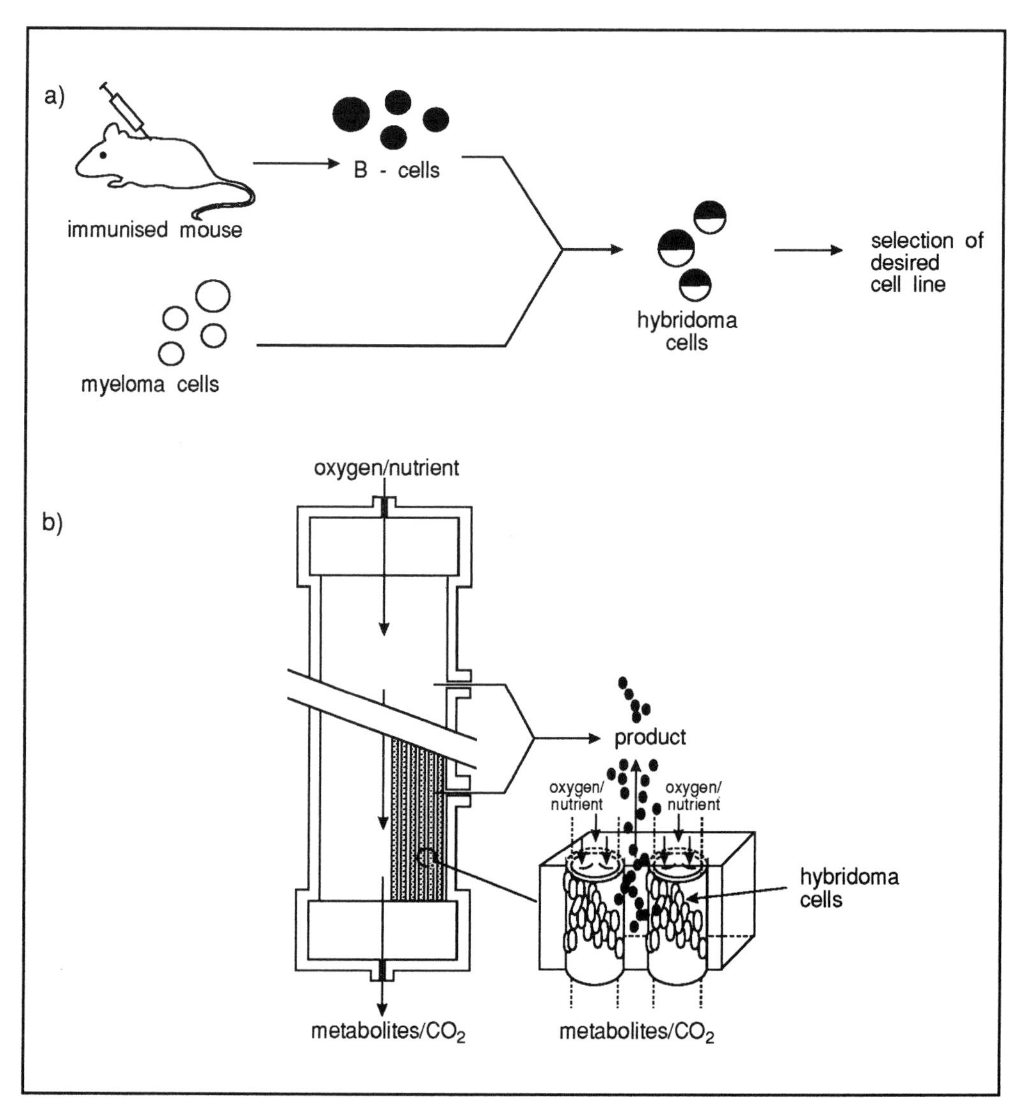

Figure 3.4 *In vitro* monoclonal antibody production a) A mouse is immunised against an antigen and the antibody producing cells (B-cells) are isolated from its spleen and fused with cancer cells (myeloma cells) to produce hybrid cells (hybridoma cells). The hybridoma cells are screened and the cells producing the desired antibody are isolated. b) The desired hybridoma cells are grown on a large scale. In b) we have illustrated a hollow fibre bioreactor used for this purpose. Nutrients, wastes, gases and metabolites are exchanged over the membranes of the fibres. Note that the desired antibody (product) is collected from the extra-capillary compartment.

For various reasons there was a need to develop *in vitro* systems to produce monoclonal antibodies on a large scale. These reasons include:

- antibody yield in mice is variable;
- murine ascitic fluids are contaminated by murine proteins and sometimes by viruses;
- *in vivo* production does not fit with the current trend to reduce the use of laboratory animals.

An example of an *in vitro* culture system is given in Figure 3.4.

We should anticipate a great extension in the use of monoclonal antibodies in animal husbandy in the future. We will see examples of how monoclonal antibodies can be used for diagnostic purposes in the final chapter. Here we will focus on their use in monitoring reproductive status.

3.6.2 Monitoring reproductive status

Several methods of monitoring reproductive status in animals are available. Some, such as rectal examination for pregnancy determination in cattle, have been used for many years. Others have only relatively recently been developed. The extent to which these methods are currently used, and the prospects of their future use by veterinarians and farmers, vary.

Π From what you have learnt about reproduction and monoclonal antibodies, can you write down a possible scheme for determining whether or not an animal is pregnant.

We would anticipate that you would suggest that monoclonal antibodies could be used to detect pregnancy specific hormones. A good example would be to measure oestrone or progesterone. In fact the concentrations of the reproductive hormones in peripheral blood and milk are extremely low, and specialised immunological techniques have to be employed to detect them. The reaction between an antigen and an antibody is specific and rapid, and can be used to produce tests sensitive enough to measure circulating hormone concentrations as low as pg ml^{-1}. The development of monoclonal antibodies has enabled the development of commercial test kits.

Recently developed tests allow detection of hormones in blood or milk within minutes, have simple visually detected endpoints, and require no specialised equipment. They are therefore ideally suited for use in the clinic or on the farm, either by the veterinarian or farmer.

From among the various tests available to measure pregnancy-specific hormones, the application of progesterone-test is outlined below.

Other similar tests include:

- PMSG-test to measure equine chorionic gonadotrophin (eCG=PMSG) in the pregnant mare;
- Oestrone sulphate-test, to test for pregnancy in the pig for example.

Measurements of progesterone have the most widespread application for fertility monitoring. Among them are:

- detection of non-pregnancy;
- checking the timing of insemination;
- checking normal ovarian function;
- assessing the stage of the cycle for embryo transfer;
- differentiation between follicular and ovarian cysts.

bovine milk progesterone test

The concentrations of progesterone in milk of dairy cows and goats reflect those in blood, making this the biological fluid of choice for these species. The bovine milk progesterone test is probably the most widely used hormonal fertility test. Typical progesterone levels in milk around the time of oestrus and insemination are shown in Figure 3.5.

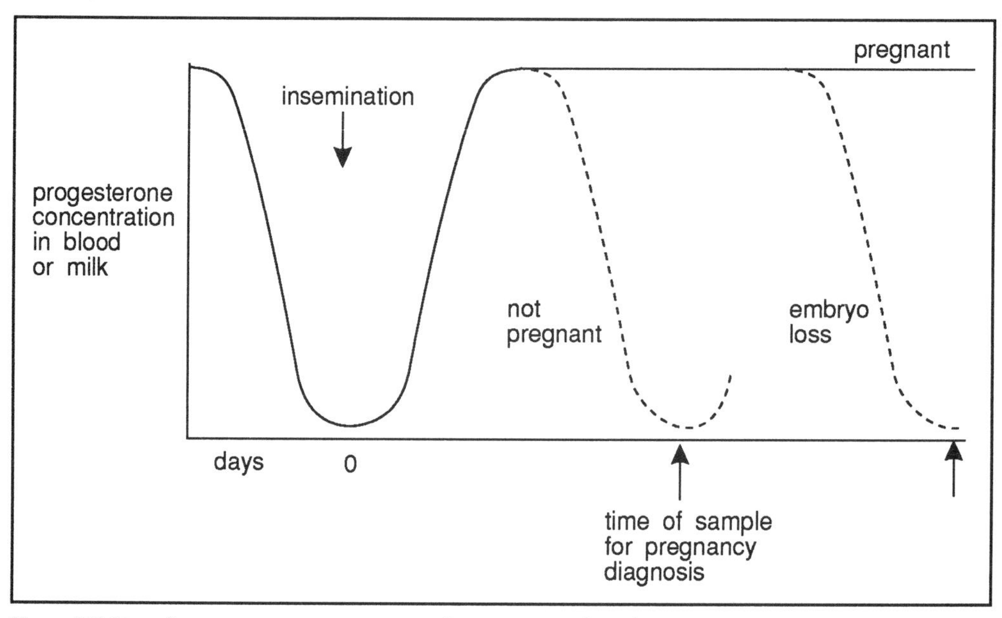

Figure 3.5 Use of progesterone measurements for pregnancy detection.

SAQ 3.3

Why is the bovine milk progesterone test now better described as a test for non-pregnancy, rather than pregnancy?

SAQ 3.4

Answer true or false to the following statements.

1) Synchronisation of oestrus and ovulation can be achieved by holding animals in an extended luteal phase by administering PGF2α.

2) A practical way of inducing oestrus in cattle is to administer exogenous gonadotrophin.

3) In the induction of superovulation in cattle using PMSG, the use of antibodies against PMSG improves the development of embryos.

4) Monoclonal antibodies against luteinising hormone (LH) would make an excellent detection system to determine whether or not an animal is pregnant.

5) Pregnancy may be terminated by prostaglandin treatment.

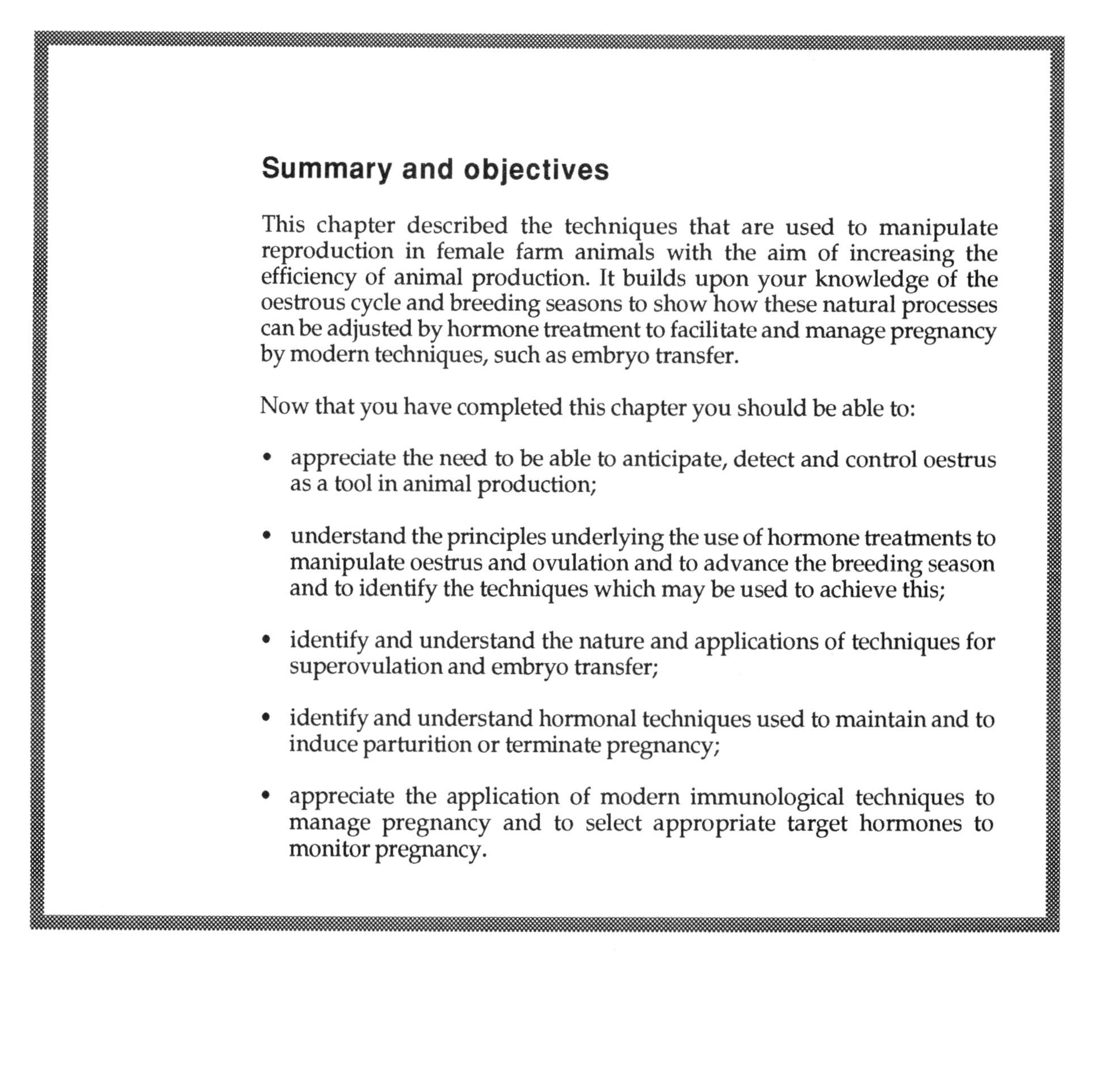

Summary and objectives

This chapter described the techniques that are used to manipulate reproduction in female farm animals with the aim of increasing the efficiency of animal production. It builds upon your knowledge of the oestrous cycle and breeding seasons to show how these natural processes can be adjusted by hormone treatment to facilitate and manage pregnancy by modern techniques, such as embryo transfer.

Now that you have completed this chapter you should be able to:

- appreciate the need to be able to anticipate, detect and control oestrus as a tool in animal production;
- understand the principles underlying the use of hormone treatments to manipulate oestrus and ovulation and to advance the breeding season and to identify the techniques which may be used to achieve this;
- identify and understand the nature and applications of techniques for superovulation and embryo transfer;
- identify and understand hormonal techniques used to maintain and to induce parturition or terminate pregnancy;
- appreciate the application of modern immunological techniques to manage pregnancy and to select appropriate target hormones to monitor pregnancy.

In vitro embryo production and manipulation

In vitro embryo production and manipulation

4.1 Introduction

The aim of animal breeding is to combine many good qualities in one animal. Since the early days of domestication of animals, Man has selected and bred from animals which had these qualities. Progress is achieved by the selection of hybrids and the breeding of multiple offspring from parents of high quality. However a single cow can yield only 6-10 calves by natural reproduction. Superovulation and embryo-transfer techniques (Chapter 3) can increase the number of good calves up to 10 per year and 100 per cow. However, the number of embryos that can be obtained by such means is still the main factor slowing progress in breeding.

Recently many new techniques have been introduced in the field of animal reproduction and breeding. Later sections will describe the following:

- production of multiple embryos *in vitro* (Section 4.2);
- asexual multiplication of embryos by splitting or nuclear transfer (Section 4.3);
- evaluation of embryo quality (Section 4.4);
- embryo sexing (Section 4.5);
- cryopreservation (Section 4.6);
- micro-injection of embryos (Section 4.7).

4.2 Production of multiple embryos *in vitro*

Although only a single follicle will ovulate for each oestrus in the cow, in most animals at least three different groups of large (10mm in diameter) and relatively large (2 10mm) follicles can be distinguished in the ovary during one oestrous cycle. In total nearly 50 to 80 follicles of 2mm diameter grow during the cycle but all except one degenerate. So during each cycle of 21 days a large number of oocytes are excluded from maturation, fertilisation and embryo development.

Π How might more of these oocytes become available for fertilisation?

As we described in Chapter 3, researchers try to maintain the development of several follicles and oocytes by injecting gonadotrophic hormones like FSH and PMSG. The result of that treatment in most cases is the ovulation of many follicles (superovulation) and, after insemination of the cow, the development of many embryos. However, superovulation procedures still suffer from highly variable yields of transferable embryos. Research needs to be done for a better understanding of what happens in an ovary after hormonal treatment. The regular ovulation of one oocyte in the cow results from the normal endocrine control as described in Chapter 2. This is influenced by many regulatory feedback mechanisms, and may not be able to tolerate the hormonal treatments used for the induction of superovulation. Individual variability of responses is also a problem. A regular yield of 4 to 6 good embryos after a hormonal treatment may be the limit that can be achieved. Therefore alternative methods have been explored. One of the most successful alternatives to superovulation is the *in vitro* maturation and fertilisation of immature bovine oocytes and their subsequent development up to the blastocyst stage.

in vitro maturation and fertilisation

Write down the major advantage that *in vitro* maturation might have over superovulation?

transvaginal ultrasound guided puncture

This procedure is attractive since immature oocytes can be collected repeatedly from the same living animal irrespective of the day of the oestrous cycle. The method recently developed for this purpose is the 'transvaginal ultrasound guided puncture' of follicles, which uses ultrasound to guide a needle via the vagina into the follicles to extract the oocytes. In this way many oocytes can be collected from a single animal. To understand this process we need to learn a little about ooctye development.

oocyte development

Primordial germ cells arise very early in the development of the female bovine embryo from the primitive endoderm. They migrate to the ventral surface of the mesonephros (primitive kidney) where the genital ridges develop. In cow embryos migration is complete by day 35 post-coitum.

Between the 48th and 150th day of pregnancy the oogonia undergo rapid mitotic divisions, reaching a peak number of about 3×10^6 germ cells. The onset of meiosis occurs between 75 and 80 days post - coitum and proceeds to the diplotene stage, whereupon it is blocked by a maturation inhibiting factor. Resumption of meiosis in oocytes only takes place in the preovulatory follicular stage about 24 h before ovulation in adult life. The population of oocytes diminishes by waves of degeneration. At birth only 1×10^5 oocytes are left in 'the stock' of primordial follicles, from which all the developing follicles subsequently emerge.

Follicular growth up to and including the antral stage (tertiary follicles) is commonly found during sexual maturity as well as immaturity. Ovulation however does not occur before the onset of puberty. Once the follicle has started to grow, it proceeds without interruption to ovulation or until it becomes atretic (incapable of ovulating and being fertilised) at some stage of the development. Only one follicle ovulates every 21 days.

What is the fate of the remaining developing follicles and their oocytes following ovulation?

All other follicles (about 80) including their oocytes, degenerate sooner or later during the cycle. However with the transvaginal puncture technique most oocytes of tertiary follicles can be collected before the degeneration of the follicles ends in the death of the oocyte. The immature oocyte can now be matured in the laboratory within a 24h culture period and subsequently fertilised.

sperm capacitation

zygotes

Ejaculated bovine sperm, however is not able to fertilise the oocytes *in vitro*. It first needs to be treated with heparin or other substances like caffeine and lysophosphatidyl-choline to become mature and 'capacitated'. Little is known about the fine processes which occur in the sperm cells before they can be classified as capacitated and capable of fertilisation. Nevertheless, in recent years, young in limited numbers have been produced in cattle by *in vitro* maturation and fertilisation. In most cases development to the blastocyst stage was reached by transferring the fertilised oocytes (zygotes) into the fallopian tube of a recipient sheep. After six days the genital tract of the ewe was flushed and the properly developed embryos were transferred to recipient cows.

More recently, embryo development *in vitro* in co-culture with tubal cells has become possible. The overall success rate of this method is about 25%. How this low percentage might relate to the developmental potency and quality of the collected oocytes is unknown. However, of all antral follicles present in the ovary at one point of the cycle, only 15 to 20% are non-atretic. This means that most oocytes collected for *in vitro* maturation come from more or less atretic follicles. Hence, this factor may limit the success of the *in vitro* embryo production procedure.

Π How might the use of atretic follicles be avoided to improve the success of embryo production?

For better production of embryos *in vitro*, oocytes should be collected from pre-antral follicles. The chance that these follicles are already atretic is very small. However, the meiotic competence of oocytes from pre-antral follicles is still questionable.

rDNA technology

There are therefore some difficulties in producing multiple embryos by collecting oocytes. Nevertheless, the production of artificially matured cattle oocytes by current technology is relevant to the development of large-scale cloning methods. We should also keep in mind that artificially fertilised cattle zygotes in the pronuclear stage have applications in the use of recombinant DNA (rDNA) technology in cattle whether for improving stock quality, for farming needs, or for producing unique human proteins for pharmacological purposes (see Chapter 5). *In vitro* production of embryos appears to have a high potential. We will move on to examine another strategy for producing multiple embryos.

4.3 Asexual multiplication of embryos by splitting or nuclear transfer

cloning

Each embryo, regardless of the method of production, is a unique combination of the genotypes of its parents. All embryos discussed thus far are different genetically. Even the embryo carrying the best inherited qualities is a single entity. Hence it should be clear why researchers have tried hard to generate more identical embryos from the single desired one: a process referred to as 'cloning'.

embryo-splitting

The method developed first was 'embryo-splitting' (Figure 4.1). Splitting divides embryos into halves, and thence into quarters. Each quarter can still develop into a viable young. Efforts to divide an embryo into eight parts have not succeeded. The number of cells then seems too small to support the development of a complete embryo. However in each nucleus of an embryo, irrespective of the number of cells, the same chromosomes are present. Each nucleus therefore has the potential to support complete embryo development. It will do so if it is transplanted into an enucleated oocyte by 'nuclear transplantation'. This technique is likely to become a widely used in the near future although its success levels are still very low.

nuclear transplantation

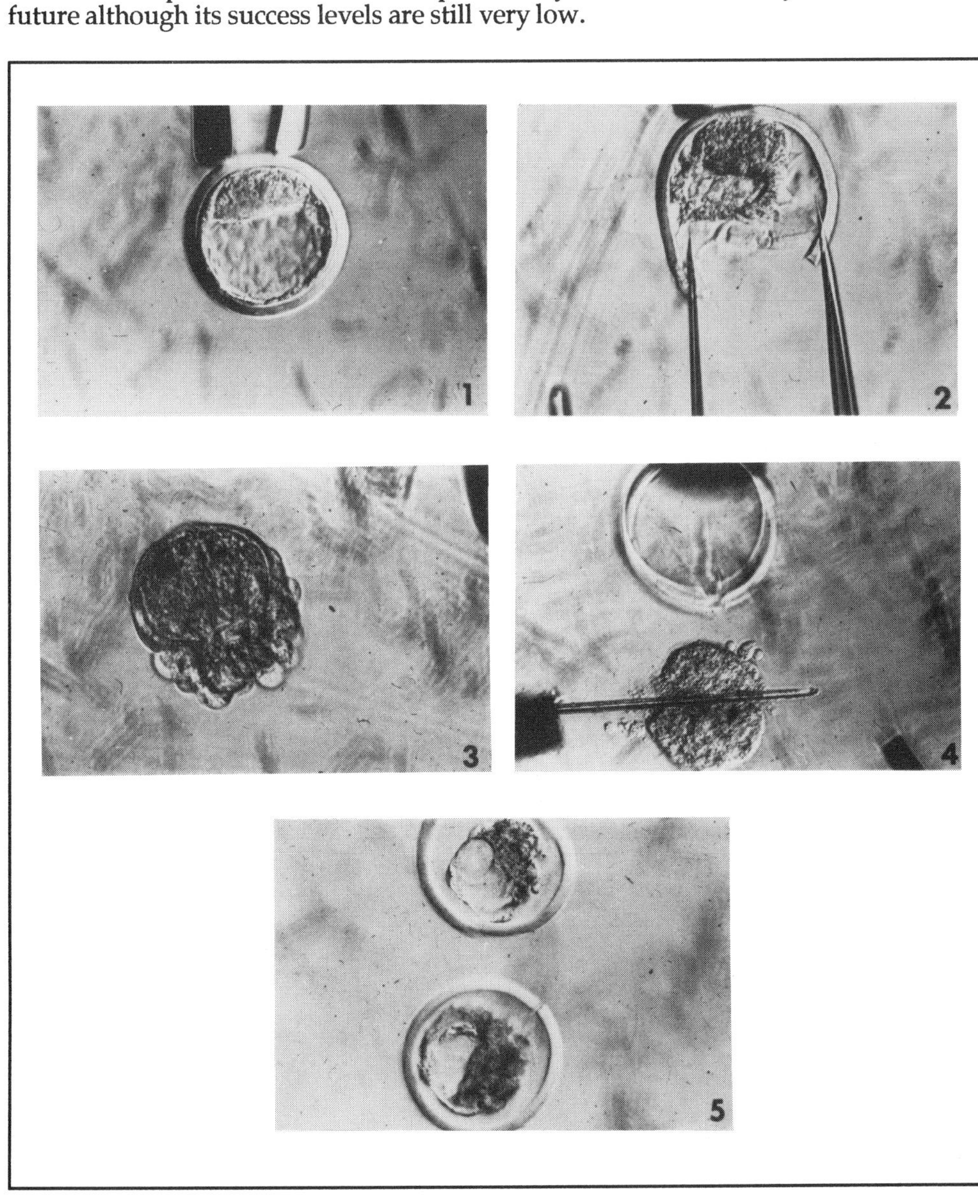

Figure 4.1 Embryo-splitting. The embryonic cell mass is first 'sucked out' of its protective membrane (1, 2 or 3) then split (4) to form the new embryonic cell masses (5).

Many factors seem to be linked to the poor viability of reconstituted embryos. Some relate to the optimal embryonic stage for obtaining nuclei. Until recently it was thought that differences in cell types and function were irreversible from the eight cell stage onwards in sheep and cattle. However, full term development has now been obtained with nuclei derived from the 32-cell stage (the morula) and even the blastula stage. Other factors are related to the sources of the cytoplasm used as recipients for the nuclei. Initially the post-fertilisation stage (zygote) was used but the cytoplasm of secondary oocytes (metaphase stage II) seems to be superior due to its ability to reprogramme nuclei from morula and blastula back to the moment of fertilisation (one cell stage).

morula

Proper activation of the oocyte and a reliable enucleation technique are important prerequisites for further normal development. Last but not least, the interactions between nuclear and recipient cytoplasm are of great importance. Corrective measures have to operate to compensate for some asynchronies between nucleus and cytoplasm. The presence of cytoskeletal inhibitors in the post-fusion medium for one hour after fusion has a positive effect on the viability of the reconstructed embryos.

A general scheme is outlined in Figure 4.2. Note that oocytes are collected from one animal and an embryo from another. The embryo is disaggregated and embryo cells are placed alongside oocytes. A pulse of electricity causes fusion of the embryo cells and the oocytes. The fused products are then re-introduced into an animal where they develop into embryos. These embryos can be used as a further supply of embryonic nuclei and the fusion steps repeated. Note that during enucleation of an occyte, some cytoplasm is also removed.

In the best experiments so far 30 to 90 cell fusions were made from one embryo but only nine embryos were reconstructed from a single one. However the enucleation procedure followed in these experiments was not very efficient. In 50% of the cell fusions the nucleus of a blastomere was fused with an oocyte with its own nucleus. These fused and reconstructed embryos did not develop. Refinements in the enucleation procedure which ensure that all blastomeres are fused to enucleated oocytes would be expected to increase greatly the efficiency of these cloning procedures. The influence of reduced cytoplasmic mass in enucleated oocytes on the developmental capacity of nuclear transfer embryos is an important area for future investigation. Developing embryos can either be used as donor embryos for a subsequent generation of nuclear transfers or be transferred to synchronous bovine recipients for development to term.

In a limited number of experiments, embryos resulting from nuclear transfer were used as donor embryos for a second generation of nuclear transfer (see Figure 4.2). From eight embryos, 84 new nuclear transfers were performed and 14 (17%) viable embryos obtained. These data indicate a reduction in the number of usable blastomeres (10.5 per embryo) when nuclear transfer embryos are used as donor embryos. The reason for a reduction in number of cells is not known but may be related to the fact that the cytoplasmic mass for the first nuclear transfer was just a half oocyte. However the reduction in cell number might have been due to the culture conditions used. A proper culture medium for bovine embryos has not yet been established.

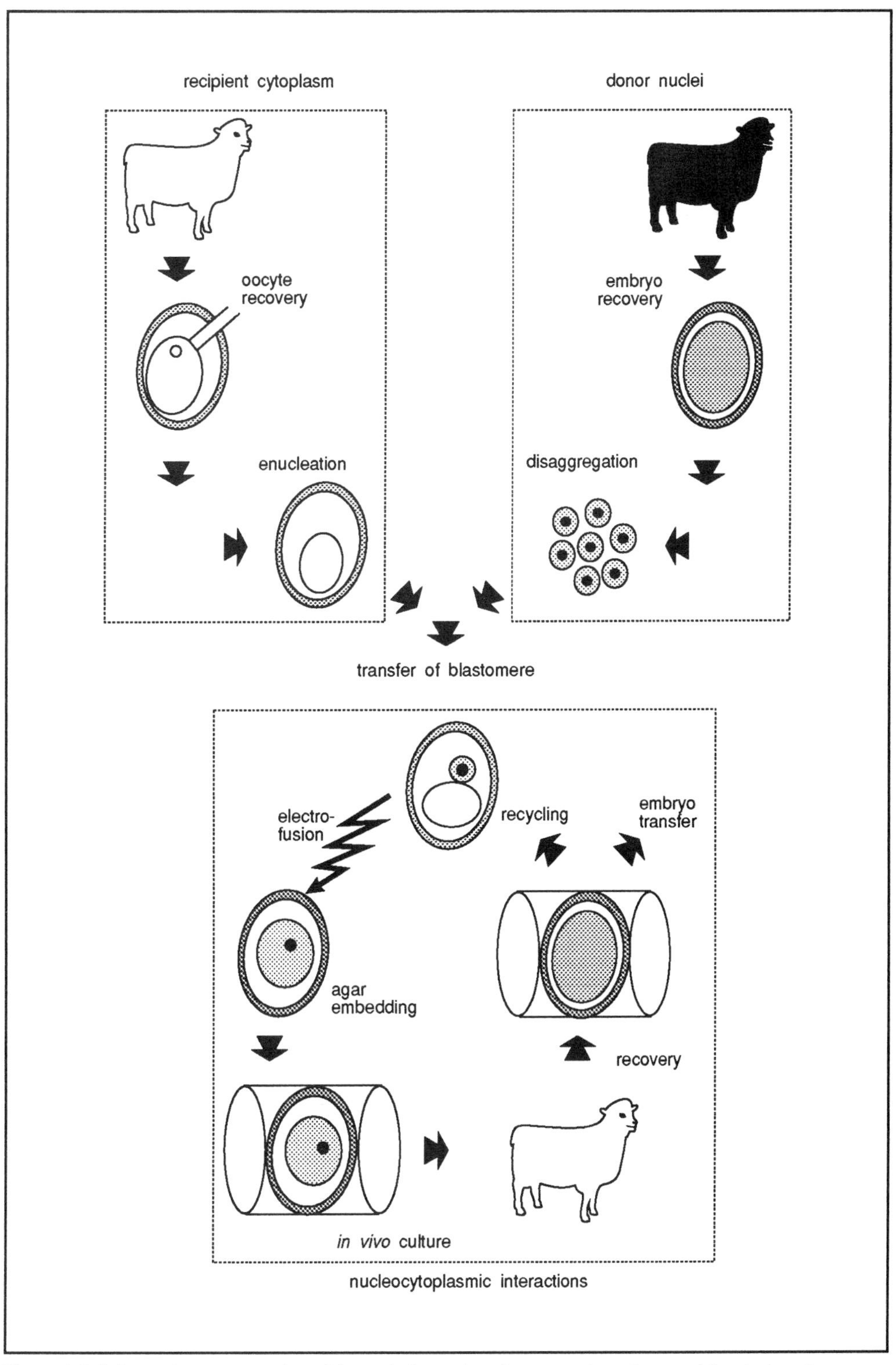

Figure 4.2 Schematic representation of the technique of nuclear transplantation used for cloning sheep embryos.

Recent data on embryo development and pregnancy rates following transfer of nuclear transfer embryos show a lower viability of the embryos than is observed normally for non-surgically collected and transferred intact embryos. Nevertheless the embryos have the same grading, ranging from excellent to poor. The reduced pregnancy rate is unexplained at this time.

pluripotent cell lines

Despite the difficulties, the commercial cloning of cattle embryos will be a reality in the near future. It will have an enormous commercial value in both the beef and dairy industry. It will have even more impact when the number of available blastomeres for nuclear transfer is unlimited. That situation will be reached when 'pluripotent cell lines' are isolated from the inner cell mass (or embryonic stem) of an embryo. Such cells show no tendency to specialised tissue formation and they develop without differentiation. Pluripotent cell lines together with large numbers of enucleated mature oocytes will potentially provide unlimited sources for the reconstruction of new embryos.

SAQ 4.1

How do embryos derived from embryo-splitting or nuclear transfer differ in their genetic composition from embryos obtained from superovulation or *in vitro* maturation techniques?

4.4 Evaluation of embryo quality

cryopreservation

Various criteria can be used to select the most suitable embryos for transfer or 'cryopreservation' (frozen embryo storage). These criteria include the stage of embryo development determined by age when harvested. It can easily be judged under a microscope. Six and seven day bovine embryos should have reached the morula and blastula stages, respectively.

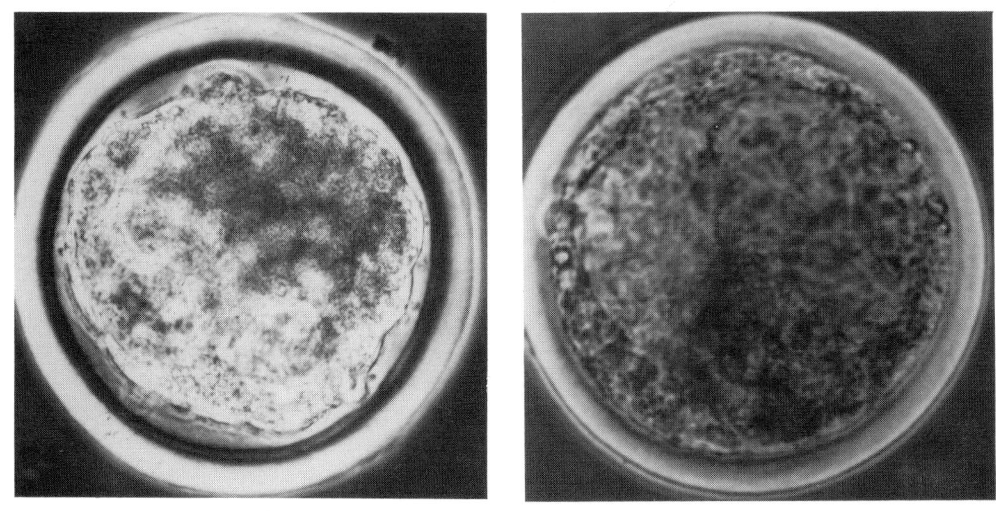

morula blastocysl

In addition to the developmental stage, the morphological appearance of the embryo can be evaluated. In rating embryo quality one should consider:

- the uniformity of blastomeres;

- the distinctness of membranes;
- the presence of vesicles;
- any damage to the zona pellucida;
- signs of degeneration and dissolution.

On these criteria the embryo can be classified as excellent, good, fair or poor. Although a close relationship has been detected between these classes and pregnancy rates, the rating on morphology alone can never be absolute. Rating quality on physiological parameters has also been tried. These include dye-exclusion tests, live-dead staining and measurement of enzymes. These tests are rather invasive. Recently a non-invasive technique has been developed. It is based on uptake of nutrients in the culture medium by the embryo. Detailed information is now available on the changing metabolic requirements of mouse embryos. Pyruvate has been shown to be an essential nutrient during the first cleavage division, with glucose being unable to support development until the 8-cell stage. It would be useful to know whether bovine embryos have similar nutrient requirements.

micro-fluorescence enzymology

An ultra-microfluorescence enzymatic method has been developed to determine the very small changes in pyruvate and glucose concentrations in microdrops of media in which the embryo has been cultured. Essentially this method uses an enzyme to catalyse a change to a substrate and this change can de detected as a fluorescence change. The technique has enabled sequential studies to be carried out on single embryos (Figure 4.3).

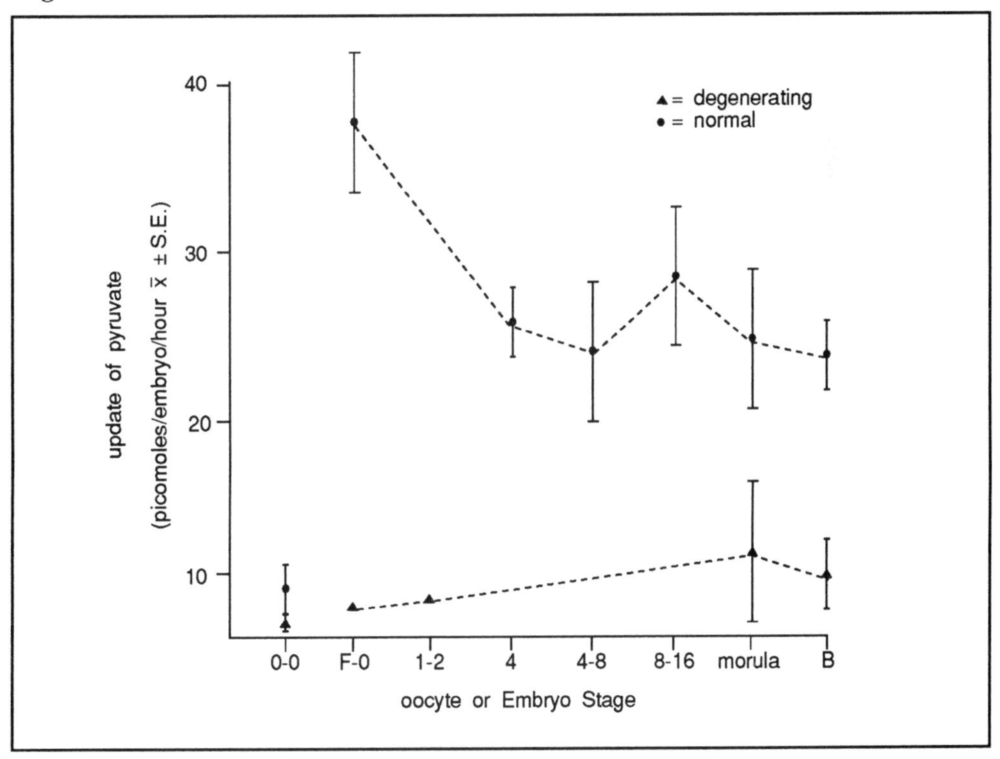

Figure 4.3 Uptake of pyruvate by human oocytes and early embryos. 0-0 = Old oocytes: F-O = fresh oocytes: 1-2, etc = embryo stage: blastomere number, B = blastocyst.

Π Is there any evidence that the pyruvate requirements of mouse and human embryos are similar? Give an application for the technique.

Yes. As with human embryos the pyruvate requirement of mouse oocytes is highest at the first cleavage division and thereafter shows some decline. The technique could be used to distinguish normally developing oocytes and embryos from degenerating ones by analysing pyruvate uptake from culture media.

4.5 Embryo sexing

immuno-fluorescence

DNA-probes

The earliest method for the determination of the sex of an embryo was based on taking a biopsy of an embryo, culturing these cells and fixing them in metaphase. Then the karyotype (chromosomes complement) was analysed. This method however is invasive and time consuming. Modern tests use 'immunofluorescence' based on the fact that male embryos have the so-called HY-antigen on the outer membrane of their cells. Monoclonal antibodies conjugated with a fluorescent vital dye will bind onto the membranes of male embryos. However non-specific binding of the antibodies on embryos of poor quality is still an unsolved problem in sexing embryos. More recently 'DNA-probes' have become available. These make use of the property that they will 'recognise' and bind to a specific sequence of nucleotides in the genome. These can therefore be used for direct binding to the genome. Although this technique is very reliable, a single blastomere still needs to be removed from the embryo.

4.6 Cryopreservation of embryos

Remarkable progress has been made in freezing and use of frozen embryos. It offers many advantages but the major one is that one can preserve all viable embryos until recipients at a matched physiological state are available. Embryos can be transported at any time or to any place where recipients are available. Freezing embryos is an essential prerequisite for a sound programme of *in vitro* embryo production. Cryopreservation offers an inexpensive and efficient way to maintain an embryo bank. Basically the procedure for freezing involves various modifications of the following key steps:

- placing the embryo in a culture medium containing a cryopreservative, such as glycerol;
- cooling the embryo to -7°C, below the freezing point of the solution;
- seeding the vial or straw containing the embryo with ice crystals using a very cold rod held away from the embryo;
- cooling to -30°C at a cooling rate of about 0.3°C min^{-1};
- finally, transferring the embryos to liquid nitrogen.

Π Write down a reason why the embryo should not be quick frozen directly by immersion in liquid nitrogen.

Quick freezing would kill the embryo through disruption of cells as ice forms. To avoid this cryopreservative or cryoprotective agents like glycerol are used in the culture medium together with slow, controlled freezing rates. Dehydration of cells occurs uniformly avoiding damage by intracellular ice crystals and membrane disruption.

Thawing of the embryo can be done relatively quickly by placing it in water at 30°C. Because the embryo will contain a high concentration of glycerol and salts in the frozen state, it will swell by osmosis upon thawing and can burst unless steps are taken to counteract this. One approach to prevent swelling damage is to transfer the embryo through a series of solutions with decreasing glycerol concentrations. Another approach is to package the embryo in a straw containing both freezing and thawing media. The freezing solution containing the embryo is separated from the thawing solution by an air space. The thawing solution contains a high concentration of sucrose, of the same osmolality as that of the freezing solution. However, sucrose is a nonpenetrating substance. When the straw has been thawed it is immediately flipped by hand to break the air bubble and move the embryo into the thawing solution.

The straw system described is one way to handle frozen embryos for use in embryo-transfer. It does have a disadvantage: the embryo cannot be examined upon thawing. Thus it is important to freeze only embryos of high quality, as they tend to survive freezing well. Frozen embryos give pregnancy rates about 10% lower than with fresh embryos.

4.7 Micro-injection of embryos

gene transfer

Revolutionary new opportunities for the modification of animal performance have been created by the development of methods for gene transfer. Currently, transfer of a gene in farm animals is achieved by direct injection of several hundred copies of a gene into the nucleus of a recently fertilised oocyte ie the zygote. Gene transfer by direct injection is quite difficult in lifestock. The reason is that bovine oocytes, for example, contain lipid droplets and cytoplasmic vesicles which conceal the nuclei. This problem can be solved by centrifuging the zygote shortly before injection. Droplets and vesicles are piled up on one side of the embryo and the nuclei become visible. Fortunately, the majority of embryos are able to develop normally after this treatment. However, the proportion of injected eggs that develop into transgenic offspring is very small, being around 1% in farm animals. The potential of gene transfer will be described in more detail in the next chapter.

SAQ 4.2

Match the technique with the appropriate function in embryo manipulation.

Technique	Function
superovulation	to provide cells into which blastomere nuclei can be transferred
enucleation	to enable sexing of embryos
microinflourescence enzymology	to fuse blastomeres and oocytes
electrofusion	to detect metabolic status of embryos
DNA probing	to produce multiple oocytes

Summary and objectives

This chapter described how various techniques can be used to obtain increased numbers of good quality offspring from suitable parent animals by augmenting natural reproductive processes, particularly at the embryo stage. Natural ovulatory mechanisms may be enhanced, or even replaced by *in vitro* techniques of oocyte maturation and fertilisation. Multiple embryos of related or identical genotypes may be produced for implantation and subsequent development, or stored frozen until required. Embryo sex and quality may also be assessed.

Now that you have completed this chapter you should be able to:

- describe the uses of superovulation, *in vitro* oocyte maturation and fertilisation techniques to yield multiple embryos of good quality;
- understand the basis of embryo-splitting and nuclear transfer techniques or cloning used to produce multiple embryos of good quality and identical genotype;
- recognise the value of non-destructive techniques to judge embryo quality and viability on morphological and physiological criteria;
- appreciate how embryos can be sexed by non-invasive means, or those involving minimal interference;
- describe the applications of embryo cryopreservation and subsequent thawing for implantation into recipient animals;
- select appropriate techniques for producing, characterising and storing embryos.

Gene transfer to a whole animal: the production of transgenic (livestock) animals

Gene transfer to a whole animal: the production of transgenic (livestock) animals

5.1 Introduction

In 1987 Peter Newmark wrote in his comments on the Nature Conference on Plant and Animal Biotechnology: "In the farms of the future, as in Orwell's 'Animal Farm', all animals will be equal, but some animals will be more equal than others". The transfer of foreign genes to the embryos of farm animals has shown that this future is becoming more and more a reality. The objectives of this chapter are to illustrate the tools needed to produce transgenic animals and to discuss some of the applications of the technology. The first successful transfer of DNA to a fertilised mouse egg was reported in the early 1980s. Since then many genes have been transferred to mice and also to livestock animals.

transgenic animals

chimaeric animal or chimaera

When foreign DNA is introduced into a fertilised egg before the egg has divided, there is a good chance that the DNA will be inherited by all of the cells of the embryo, including the germ cell line. Hence, after the development of an adult animal the foreign DNA can be transmitted to any progeny. Animals that have been manipulated in this way are referred to as transgenic animals. Introduction of foreign DNA into a later stage of the developing embryo results in a chimaeric animal, or simply a chimaera. By chimaera we mean an organism that contains tissues made up of cells containing at least two different sets of genes. Thus these genetic mosaics will contain tissues of more than one distinct genetic type.

SAQ 5.1

Will animals bred from chimaeras include the foreign DNA in the genotype of their progeny?

5.2 Gene transfer methods and target cells

5.2.1 Micro-injection into pronuclei

The ultimate goal of gene transfer to a whole animal is to obtain transgenic offspring. Therefore the method chosen to introduce DNA is micro-injection of a solution containing DNA into a fertilised egg which has not yet divided, or zygote.

Fertilised eggs should be examined by microscopy to identify among them those which contain the two pronuclei. One micro-injection needle is used to hold such a fertilised egg in place, and a second extremely fine needle is used to inject a few picolitres (pl) of a solution containing 200-2000 copies of DNA into the male pronucleus (Figure 5.1). The male pronucleus in the mouse is larger than the female and therefore easier to use for injection. Special equipment has been developed which makes it now possible to inject about 100 fertilised mouse eggs per hour.

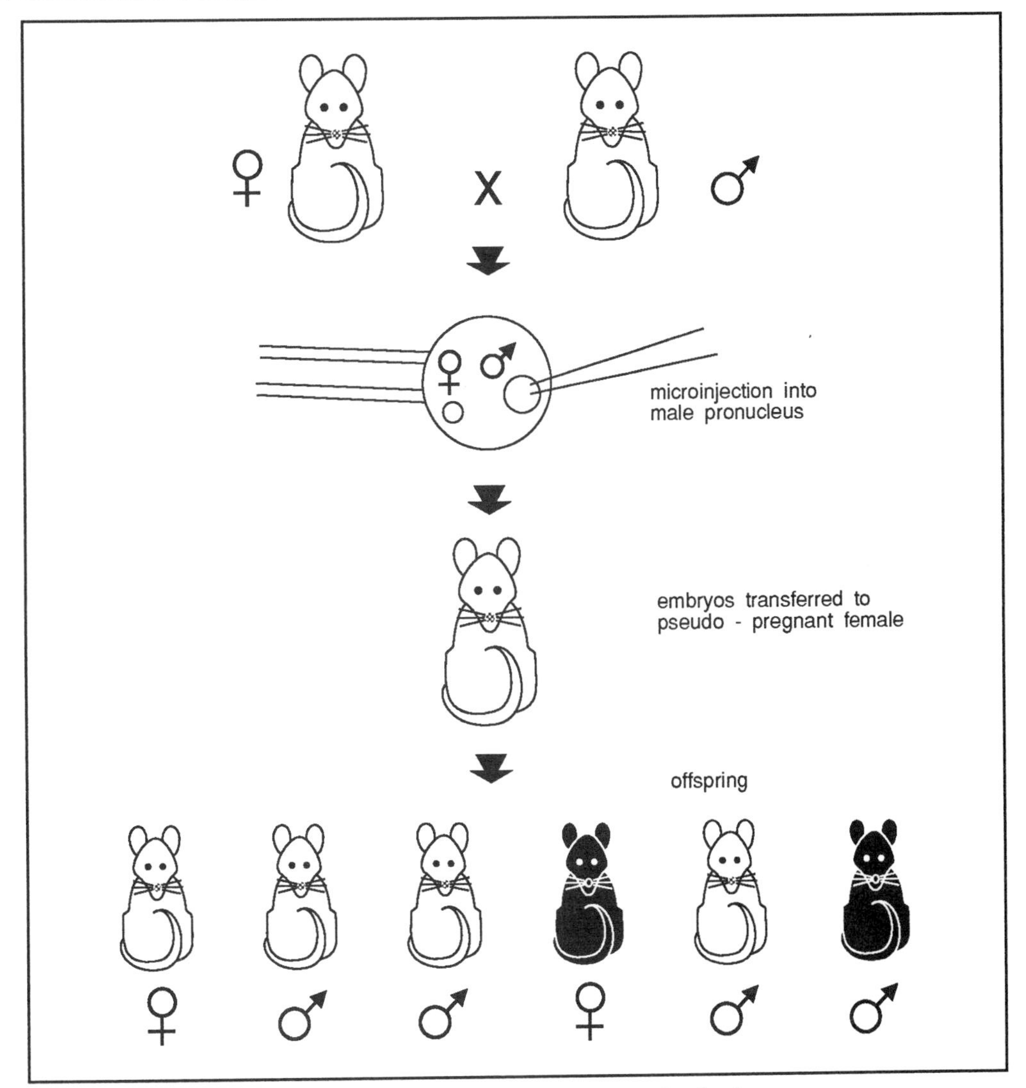

Figure 5.1 Micro-injection of mouse zygotes. Black mice are transgenic offspring.

Unlike injection of fertilised mouse eggs, manipulation of fertilised eggs of livestock animals is a tedious job. A substantial proportion of these eggs do not contain pronuclei, and even more problematic is the fact that their cytoplasm is opaque and not translucent like that of mouse eggs. Visualisation of the pronuclei of livestock animal eggs has been achieved by differential interference contrast microscopy, by centrifugation and by a combination of both these methods.

pseudo-pregnancy

In the mouse micro-injected eggs are subsequently transferred to a pseudopregnant recipient on the same day, or following incubation *in vitro* overnight or even longer, by which time the egg has divided at least once. Pseudopregnancy results from an infertile mating and prolongs the luteal phase of the oestrous cycle to favour implantation of the blastocyst.

Using the procedure described above, an overall transgenesis efficiency of about 1-10-% (based on the total number of injected eggs) has been reached in the mouse system.

Partly due to the technical problems described above the efficiency for production of transgenic livestock animals is at least a factor of 10 lower then that achieved with mice.

5.2.2 Viral infection of early embryos

retroviral vector

A very efficient method of introducing foreign DNA into a host cell is to use a retroviral vector. Upon entry retroviruses introduce their RNA genome into eukaryotic cells, and one of the key steps in their multiplication is the reverse transcription of their genome. The resulting single-stranded DNA copy is made double-stranded and is subsequently integrated as a single copy into the host chromosome. Any gene that has been cloned in such a retroviral vector can thus be, in principle, transferred to a eukaryotic cell.

No special equipment (like micromanipulators) is necessary to carry out such a DNA transfer experiment. Cells of the mouse embryo can be infected with a murine retrovirus either by co-culturing 4-8 cell embryos with virus - producer cells or by injecting the virus directly into the blastocyst. Virus-producer cells are cells that are infected by the virus in which viruses are produced and released.

Only when foreign DNA is integrated into embryonic cells which during development contribute to the germ line will the DNA be transmitted as a Mendelian trait. Furthermore, since integration takes place at different sites in different cells, the offspring are often chimaeras or genetic mosaics. Outbreeding of the offspring will then be necessary to obtain pure lines.

5.2.3 Transfer of transfected cells to early embryo

embryonic stem cells

As we discussed in Chapter 4, embryonic stem (ES) cells are pluripotent cells, ie when introduced into an 8-cell stage embryo ES cells contribute to the development of all tissues in the embryo. For this reason, ES cells are excellent routes for the introduction of foreign DNA into a whole animal.

transfection

For transfection or introduction of exogenous donor DNA into ES cells, microinjection and retroviral vectors can be used. Following transfection of ES cells transformed cells can be selected *in vitro* for the expression or insertion of the input DNA and then subsequently be used for transfer to the early embryo.

Write down a reason why the efficiency of production of transgenic livestock is likely to be lower.

You probably indicated that the opacity of zygote makes it difficult to see the male pro-nucleus and that the failure of some systems to produce such pro-nuclei as the likely reasons.

The gene transfer methods described so far in this section have been applied in the transgenic mouse system. Transgenic livestock animals have been obtained solely by the micro-injection technique. Since the ES system for livestock animals is under development, it is reasonable to expect that transfection of livestock ES cells followed by fusion with an enucleated mature oocyte will become possible in the near future.

homologous recombination

An important application of the ES system is targeted insertion of a transgene in place of an endogenous gene by homologous recombination. Homologous recombination is a rare event, and recombinants can only be identified against the background of random insertions using very sensitive screening and/or selection systems. The enormous potential of this approach should be apparent. In principle it should become possible to generate animals of any desired genotype. Genes could be mutated, endogenous genes could be inactivated and be replaced by mutant genes or by an analogous gene from a different species, and defective genes could be corrected at will.

SAQ 5.2

Contrast the processes of micro-injection, retroviral vectors and ES cell transfection for their likely success in producing transgenic animals.

5.3 The design of a transfer vector

Π Gene expression is controlled at several levels. Make a list of them?

Control may be exerted over the rates of transcription and translation of mRNA processing and breakdown of RNA, and protein modification and degradation. Thus on your list you might have included:

- accessibility of the DNA to RNA polymerase. DNA condensed in histones (heterochromatin) is not transcribed. Likewise, the binding of repressor molecules may prevent transcription;
- before an RNA is translated, it often has to be 'processed' (ie introns removed by splicing). This process may be quick or slow, and thus govern the extent and speed with which a gene is expressed;
- mRNA stability;
- post-translational modification of the protein;
- protein breakdown.

It is impossible here to discuss in detail the state of the art regarding the control of gene expression. However, one example has been chosen, which illustrates some of the regulating sequences that control gene expression. (For further general discussion of gene regulation we recommend the BIOTOL text 'Genome Management in Eukaryotes').

Suppose we wish to obtain a transgenic mouse which secretes in its mammary gland the protein, human serum albumin (HSA). The following sections explain what genetic information should be included in the transfer DNA to achieve this goal.

5.3.1 The gene encoding HSA

This can be either the intact HSA gene or a cDNA copy of its mature mRNA. The latter will only contain the exon sequences which are fused together during processing, eg by splicing of the primary HSA transcript.

5.3.2 Regulatory sequences

transcription control elements

Many genes are only expressed in specific tissues, and this is regulated at the level of transcription. Transcription of mRNAs is carried out by the enzyme, DNA-dependent RNA polymerase type II (pol II). Some of its transcription control elements, eg promoter sequences, are schematically shown in Figure 5.2.

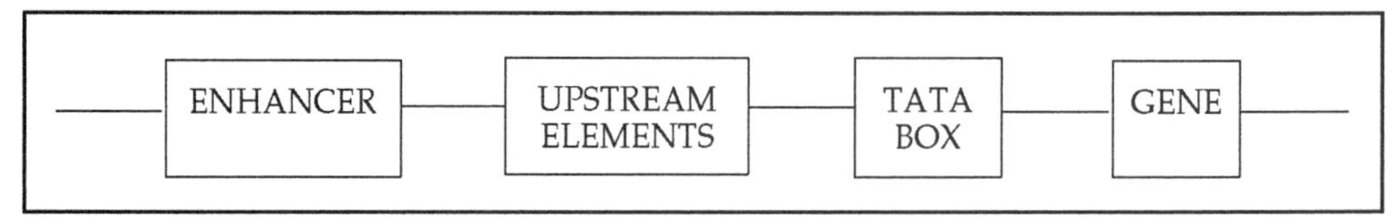

Figure 5.2 Key elements of a hybrid HSA gene to give HSA secretion by the mammary gland of a transgenic animal. (see text for details).

TATA box

upstream elements

Most pol II promoters contain an A+T rich region, the so-called TATA box. The TATA box is located just upstream of the transcription initiation site. About 80-100 base pairs upstream of the TATA box, the upstream elements are located consisting of general elements, the CCAAT and the GC boxes, or of elements which are specific to a certain promoter. Finally, at variable distances either upstream or downstream from the TATA box, or sometimes even within the transcribed region, enhancer sequences can be found.

enhancers

Enhancers stimulate the transcription often in an orientation-independent and position-independent way. In some cases, enhancers function only in specific cells or only exhibit activity upon induction by specific signals, for example, steroid hormones. The ability of some enhancers to determine tissue specificity can be used to obtain tissue-specific expression of a transgene.

Π Since the objective is to obtain secretion of HSA in the 'milk' what type of enhancer is likely to be appropriate?

To obtain mammary gland-specific expression, which in addition is induced only during lactation, the control elements for the expression of a milk protein should be used. The most abundant whey protein in mouse milk is the whey acidic protein (WAP). During lactation the level of WAP-specific RNA increases enormously, and it has been established that this induction is due to steroid and peptide hormones. The putative regulatory elements involved in the expression of WAP have been identified and potentially can be used to direct HSA expression in the mouse mammary gland (Figure 5.3).

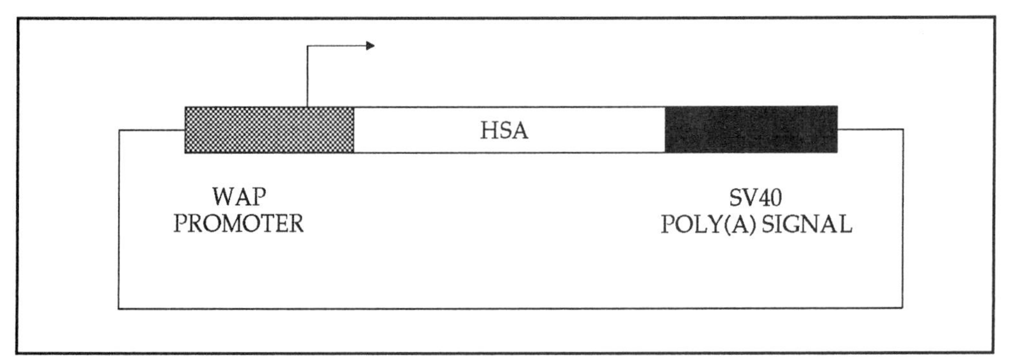

Figure 5.3 The regulatory elements for the expression of WAP linked to the HSA gene, with an SV40 polyadenylation/termination signal. The arrow indicates the start point and direction of transcription. (see text for details).

The regulatory section of WAP (the WAP promoter) can be identified by using the cDNA for WAP mRNA to identify DNA restriction enzyme fragments which carry the WAP structural and promoter gene portions by using DNA:DNA hybridisation. The promoter can be isolated from this DNA fragment.

5.3.3 Transcription and polyadenylation signals

To obtain a functional mRNA transcript, signals for termination and polyadenylation should be included in the transfer vector. These could be either the HSA signals if the complete HSA gene is used, or heterologous termination and polyadenylation signals if a cDNA of the HSA mRNA is used. A well-defined and often used polyadenylation/termination signal is derived from the small t-antigen gene of Simian Virus SV40 (Figure 5.3).

5.3.4 Protein maturation and secretion signals

signal sequence

Since HSA is a secretory protein, it is likely that the secretion signal(s) will also function in the mammary gland epithelial cells. To become translocated across the membrane of the endoplasmic reticulum (ER) a protein should contain a hydrophobic signal sequence. For most secretory proteins the signal sequence is located at the extreme N-terminus. After translocation across the ER membrane the signal sequence is cleaved off by signal peptidase, and if a stop-transfer sequence is not contained within the protein it will end up in the lumen of the ER. Secreted proteins are subsequently transported in vesicles which fuse with the plasma membrane, resulting in secretion of their contents. (This process is described in detail in the BIOTOL text 'The Infrastructure and Activities of Cells').

To obtain secretion of a cytoplasmic protein its coding region should be fused, in frame, with the N-terminal signal sequence encoding region of any secretory protein gene. In general, a membrane-bound protein can become a secretory protein if the region encoding the stop-transfer sequence eg the hydrophobic membrane anchor is deleted.

Π Consider the following stylised proteins.

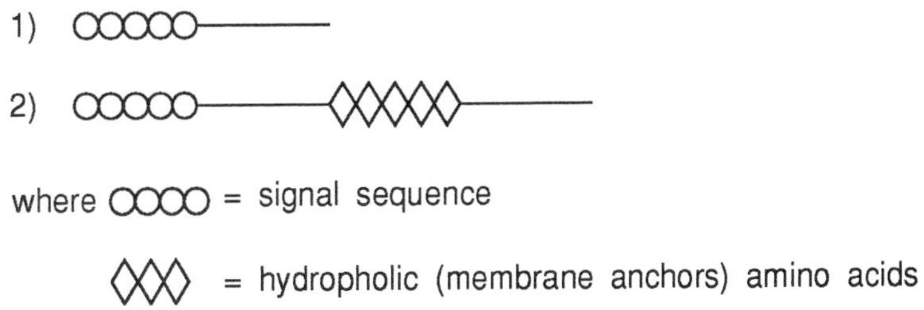

1) OOOOO———

2) OOOOO———◇◇◇◇◇———

where OOOO = signal sequence

◇◇◇ = hydropholic (membrane anchors) amino acids

Draw what would happen to these proteins once they are made inside the cytoplasm.

For protein 1) you should have come to the conclusion that it would have been secreted and drawn the sequence.

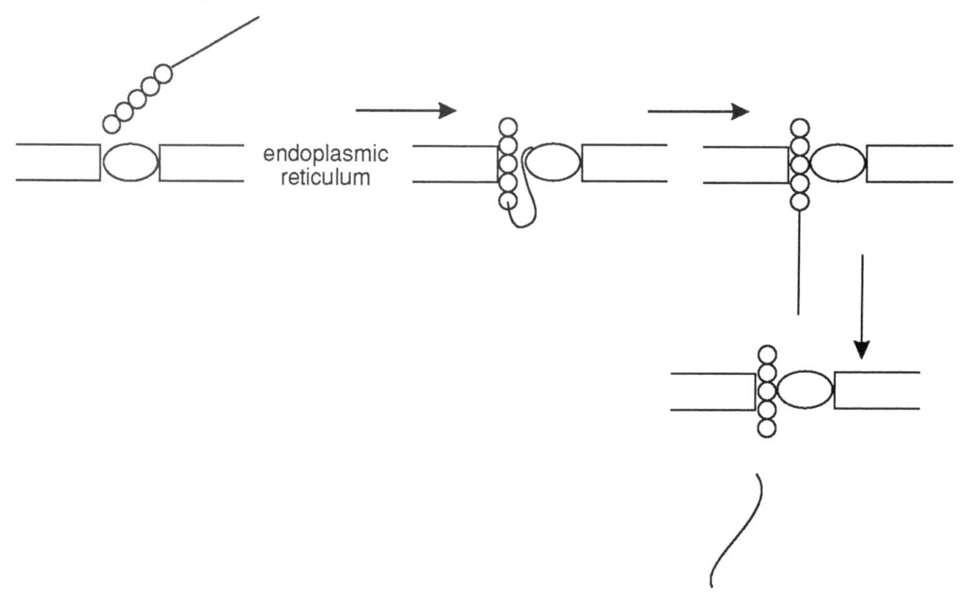

For protein 2) you should have concluded that it would have remained in the membrane. Thus:

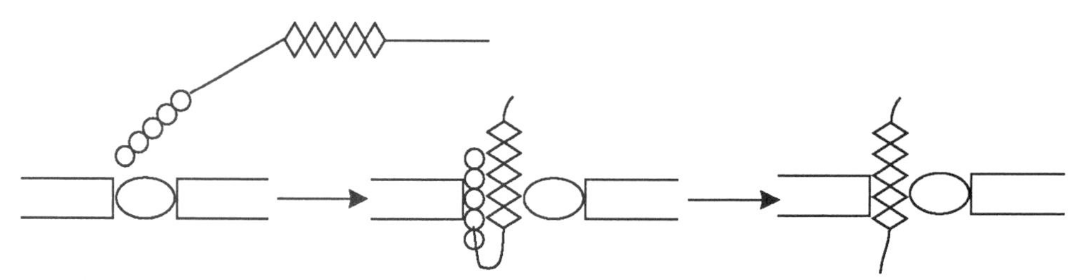

However, for many types of proteins the situation is more complex. Folding, oligomerisation, acylation, glycosylation and specific protein domains all contribute to the transport and intracellular routing of proteins, including secretion. Acylation and the type of sugars which are added to a glycoprotein are cell-dependent. If these kinds of post-translational modifications are important for the biological activity of the protein, then expression of a transgene per se does not guarantee success.

The hybrid HSA gene which is needed to obtain mammary gland expression of HSA in a transgenic mouse is summarised in Figure 5.2. For amplification in *E. coli* the hybrid gene is cloned into a plasmid, lambda or cosmid vector. The DNA for micro-injection is isolated from the bacteria, purified and either micro-injected directly or after removal of prokaryotic sequences which have been shown to inhibit the expression of the transgene significantly.

SAQ 5.3

Using the techniques and stages listed below put them in order to describe the sequence of events that need to be followed to enable the production of transgenic mouse that carries and expresses a desired gene (X) from a cat in its liver.

- isolate multiple copies of the construct from the bacteria;
- isolate $mRNA_x$ from the cat;
- attach a poly A tail to the $cDNA_x$;
- use a cDNA against a mouse liver specific gene product to enable the identification and isolation of a mouse liver specific promoter (LS-promoter);
- insert an LS-promoter-$cDNA_x$-poly A tail construct into a bacterial vector;
- inject LS-promoter-$cDNA_x$-poly A tail construct into mouse pronuclei;
- infect a bacterial culture with a vector carrying an LS-promoter-$cDNA_x$-poly A tail construct;
- make a cDNA copy of $mRNA_x$;
- attach the LS-promoter to a $cDNA_x$-poly A tail construct.

5.4 Production of transgenic livestock and its applications

The extrapolation of the transgenic mouse technology to farm animals has not been as straightforward and successful as expected. The efficiency of transgenesis has been rather low. Furthermore, livestock animals are expensive to purchase and maintain. The age of sexual maturity varies from 25-32 weeks in pigs, to 52-65 weeks for cattle. In contrast, mice can be used for breeding at an age of 6 weeks. Hence, transgenesis of livestock is a time-consuming technology. Nevertheless, several genes have been tested, and these data clearly illustrate that livestock transgenesis has become feasible. Although it is a very expensive technology, with increasing knowledge of the livestock embryology and the control of gene expression, it will certainly become a new area in biotechnology.

Π List as many potential applications of transgenic methodology in livestock animals as you can.

You probably produced an impressive list. Here we will describe the major areas of potential application.

Possible applications of the transgenic methodology are:

- The production of a large amount of specific proteins in farm animals. These could be used for: therapy, for example tissue-plasminogen activator (tpa), blood-clotting factors VIII and IX, human serum albumin; sub-unit vaccine development.

In such cases the transgenic livestock animal can be considered as a bioreactor. According to Clark *et al.* (*Trends in Biotechnology* 1987;5:20-25) the world's requirement for factor IX (with a key role in the blood clotting cascade) is about 1 kg of purified protein per year, with a value of $25,000-150,000 per gram. If factor IX expression in the mammary gland of sheep could be achieved to the same level as endogenous milk proteins, only a small flock of lactating ewes would be needed to supply the world's needs of factor IX.

- Change or improvement in the composition of milk. Hence bovine milk could mimic human milk more closely or become lactose-deficient. The latter would be important to a large number of people who are intolerant to lactose.

- The introduction of disease-resistance genes. For several economically important infectious diseases, including several enteric infections, no vaccines are available. If for a specific virus the protective immunoglobulin(s) has (have) been defined and the gene(s) cloned, it should be possible to produce transgenic animals that express virus-specific immunoglobulins in their mammary gland. During lactation the protective immunity will then be transmitted to the litter. An alternative approach would be to introduce genes encoding viral antigens, viral antisense RNA or virus-specific ribozymes which inhibit viral replication, into the germ line. Antisense RNA is RNA that has a nucleotide sequence that is complementary to normal mRNA. Thus it is capable of hybridising (forming a double strand) with normal mRNA. Such doubted stranded RNA is not translated. Thus antisense RNA blocks the production of the corresponding protein.

In tissue culture the antiviral activity of antisense RNA and of viral surface and coat proteins has been demonstrated. Transgenic plants have been produced expressing a viral coat protein which makes the plant fully resistant to infection by the corresponding virus. The use of ribozymes as an antiviral agent has yet to be demonstrated.

Of course many other applications are feasible, but all of them, including those discussed above, depend largely on capital investment. It is obvious that the development of transgenic livestock animals can only be achieved if many different technologies from embryology, cell biology, physiology, molecular biology, etc are brought together successfully.

5.5 Ethics of gene transfer

We have already raised the issue of the ethics of carrying out genetic manipulations with animals (Chapter 1). We do not intend to try to persuade, the reader of the virtues or otherwise of gene manipulation using animals. This chapter has described the techniques that can be employed to alter the genomes of animals and has explained the potential application of these techniques to producing valuable pharmaceuticals, improving the food value of livestock and of producing disease resistance in animals. Whether-or-not the reader believes mankind is justified in carrying out such manipulations is a matter for personal judgement.

Summary and objectives

This chapter described the procedures by which transgenic animals can be produced and gave examples of the applications of the methodology. Various techniques originally developed in the mouse system were described and their applications to livestock animals briefly assessed. The main design features of a transfer vector intended to give suitable gene expression in a recipient animal were developed in a typical example. Finally the applications of transgenic techniques in livestock were briefly reviewed.

Now that you have completed this chapter you should be able to:

- describe the difference between transgenic and chimaeric animals;
- appreciate the basic features, advantages and problems of the methods used to transfer genetic material between species;
- describe the basic design features of a typical transgene that will be expressed appropriately in the host animal;
- put into sequence the steps needed to produce transgenic animals using recombinant DNA technology;
- realise some of the problems but also the enormous potential of transgenic methodology applied to livestock animals.

Somatotropins in animal production

Somatotropins in animal production

6.1 Somatotropin - the molecule

growth hormone or somatotropin

somatotrophs

Growth hormone or somatotropin (GH or ST respectively) is a protein hormone synthesised in the pituitary gland of nearly all vertebrate groups. In the anterior part of this gland several protein hormones are produced, each by a distinct type of cell. Hence ST is produced in the 'somatotrophs' and found stored within these cells in secretory granules.

ST comprises a single polypeptide chain of 190 or 191 amino acids, containing two disulphide bridges (Figure 6.1). The protein has a molecular weight of 22 kD. Recently (1987) the three-dimensional structure of recombinant DNA porcine-ST has been described.

As a consequence of evolutionary divergence of the ST-gene, there are considerable differences between STs obtained from different animal groups, in both amino acid sequence and biological properties.

Although for any one species it is normal to refer to ST as a single component, in many species the hormone is known to exist of different forms. ST isolated from anterior pituitaries usually shows microheterogeneity, and also dimerisation and oligomerisation products are found (Figure 6.2). The biological significance of these different forms of ST being produced remains unclear.

The amino acid sequences for STs of at least 20 species have been reported. These were determined by protein sequencing techniques as well as recombinant-DNA techniques (deduction of the primary structure from the corresponding cDNA/mRNA nucleotide sequence). The sequences for all major farm animal species are now known (cow/bovine; pig/porcine; sheep/ovine; horse/equine; and chicken). The sequences of non-primate mammalian STs are very similar. However, compared with the sequence of human-ST the difference is marked. The homology between human- and bovine-ST for instance, amounts to only 65 per cent. As a consequence bST has no biological activity in man.

prolactin and placental lactogen

ST belongs to a family of structurally related protein hormones. Other members of this family are prolactin, which is also synthesised in the anterior pituitary, and placental lactogen found in the placenta. However, the physiological role of the latter is unclear. Prolactin has numerous actions on the mammary gland. Owing to their structural similarities, ST, prolactin and placental lactogen show overlapping activities in various *in vitro* test systems.

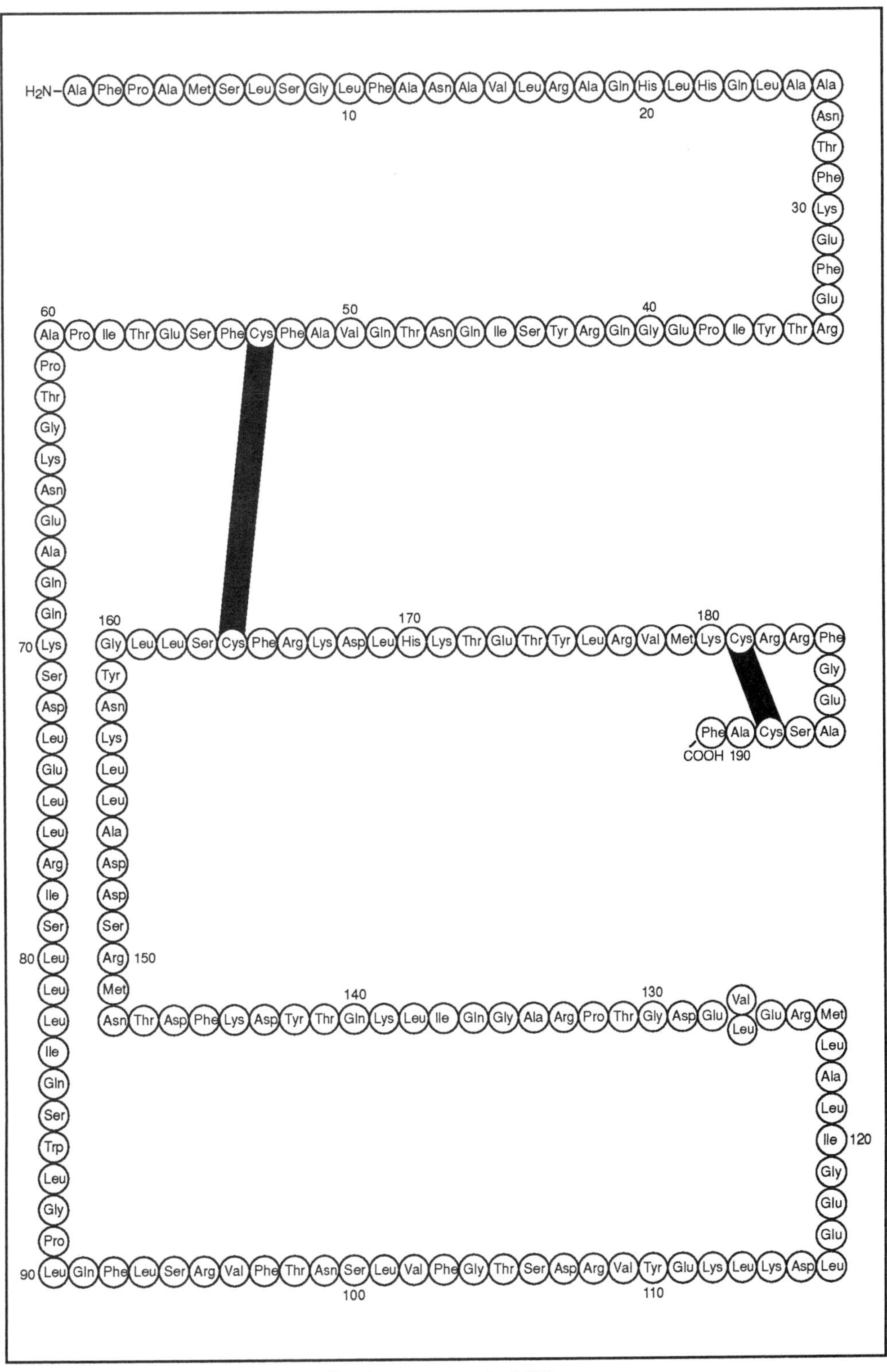

Figure 6.1 The amino acid sequence of bovine somatotropin (bST).

somatogenic, lactogenic and galactopoietic actions

However, within a given species each hormone will have a quite distinct biological role, defined by changing levels of the hormone in the circulation and by variation in the quantity, affinity and specificity of receptors in the target tissues. Thus human somatotropin has somatogenic (stimulating body growth), and prolactin has lactogenic (milk secretory) biological actions in Man. Human-ST is also biologically active in other primate and mammalian non-primates species. Bovine-ST has both somatogenic and galactopoietic (enhancing established milk secretion) effects; bovine prolactin (bPRL) however, has no effect on the cow's milk secretion during the established lactation. bST has no biological effects in primates (including Man).

	1	2	3	4	5	127	190 191
a.	Ala-	Phe-	Pro-	Ala-	Met-....	Leu/Val....	Ala-Phe
b.		Phe-	Pro-	Ala-	Met-....	"	"
c.				Ala-	Met-....	"	"
d.					Met-....	"	"

Figure 6.2 Natural variants of bST. N-terminal heterogeneity is found for bST where an Ala residue is found to precede the Phe in some of the pituitary bST (a, b). This is the result of differential removal of the signal peptide (see Section 6.6.1) rather than genetic polymorphism. Deletion derivatives c and d are minor constituents. Allelic forms are found for position 127: Leu and Val are found in about equal proportions.

6.1.1 Receptors

hormone receptors

For hormones to function, highly specific hormone receptors must be situated on or in cells of the target tissues and organs. ST-receptors are cell membrane-bound and they have been identified on many cell types (liver, striated muscle, adipose tissue, cartilage, and connective tissues). The interaction of ST with receptor binding sites on the cell surface is the first step in the biological action of ST, but little is known of the subsequent steps in this cascade. Also, the ST-receptor has been found to show heterogeneity. Recently, the primary protein structures of the ST-receptor of the liver of man and rabbit have been determined: the protein backbone contains 620 amino acid residues, and is glycosylated at several sites.

bioassay

radioreceptor assay

radioimmuno-assay

Actual measurement of ST bioactivity is done by bioassay, which makes use of known biological actions of the hormone (eg stimulation of longitudinal bone growth or gain in live weight of pituitary - deprived rats). Quantification of ST became more sophisticated with the introduction of the radioreceptor assay and radioimmunoassay. These methods respectively make use of the binding properties of purified ST-receptor preparations and specific antibodies raised against the ST molecule respectively.

SAQ 6.1

Explain how these three assay techniques are based on completely different aspects of the hormone.

6.2 Regulation of somatotropin production and release

The processes of ST biosynthesis and its release from the pituitary into the bloodstream are mainly controlled by two neurohormones. These neuropeptides are produced in the hypothalamus of the brain and reach the anterior lobe of the pituitary via a portal system of blood vessels.

growth hormone releasing hormone or GHRH

Growth hormone releasing hormone (or factor), GHRH, is a 44 amino acid peptide, which stimulates ST synthesis and release. A high degree of homology has been found between the amino acid sequences of GHRHs isolated from the hypothalami of man, cow, pig, sheep and goat (Figure 6.3). Note in this figure we have used the one-letter system of labelling amino acids. If you are unfamiliar with these, we have provided a list in an appendix at the end of this book.

```
                5    10   15   20   25   30   35   40

hGHRH     YADAIFTNSYRKVLGQLSARKLLQDIMSRQQGESNQERGARARL - COOH
                                           *   *   *
pGHRH     YADAIFTNSYRKVLGQLSARKLLQDIMSRQQGERNQEQGARVRL - COOH
                                     *     *   *  **
b,cGHRH   YADAIFTNSYRKVLGQLSARKLLQDIMNRQQGERNQEQGAKVRL - COOH
                      *              *     *   *  **
oGHRH     YADAIFTNSYRKILGQLSARKLLQDIMNRQQGERNQEQGAKVRL - COOH
          ____________________________  _________________
              intrinsic activity        species specificity and potency
```

Figure 6.3 Primary structures of mammalian hypothalamic GHRHs. The one-letter system is used to indicate amino acid residues. Bovine- and caprine (c, goat) structures are identical. Sequence substitutions for porcine, bovine/caprine- and ovine- relative to human-GHRH are denoted by asterisks. The N-terminal part of the GHRH peptide possesses the intrinsic ST-releasing activity while the C-terminal region of the molecule contains species specificity and potency.

Π What evidence is there in Figure 6.3 for the C-terminal region of GHRH determining species specificity?

Nearly all the sequence substitutions for amino acid residues are found in this region of the peptide.

somatostatin or SS

The other neurohormone or hypothalamic peptide is an ST-release inhibiting factor named somatostatin (SS), a small 14 amino acid peptide (Figure 6.4).

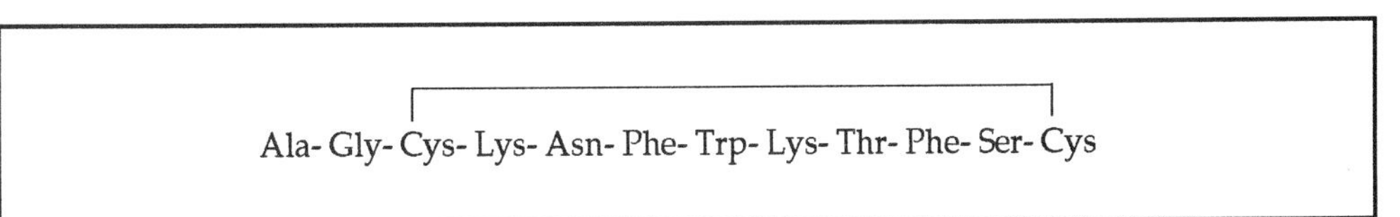

Figure 6.4 Primary structure of somatostatin (SS). Somatostatin is found in both a linear and a cyclic (disulphide) form, both of which appear to be equipotent. Besides inhibiting the release of ST, somatostatin is known to inhibit the release of many other hormones including insulin and thyroid stimulating hormone. Moreover it controls the release of many gut hormones.

Both these peptides in turn are controlled by the central nervous system, the hypothalamus being the centre of neuroendocrine integration. Many stimuli of metabolic origin are integrated at this level to influence ST-release. Feedback systems also operate at different levels to control ST-release. Both SS and GHRH feedback to inhibit their own release and also positively feedback to stimulate secretion of their opposing neuropeptides. Also ST itself may act centrally to inhibit GHRH or stimulate

SS-release. Finally it seems that circulating IGF-I (Insulin-like Growth Factor, see Section 6.2.2) feedbacks on both the hypothalamus (inhibition of GHRH synthesis) and the pituitary gland, although SS is a much more effective inhibitor of total ST release.

Π Now re-read the last section and attempt to construct a diagram to show how ST secretion from the pituitary is regulated. When complete it should bear some resemblance to Figure 6.5 though it would be unreasonable to expect that you will have included all of the elements. Your drawing may have looked a little different.

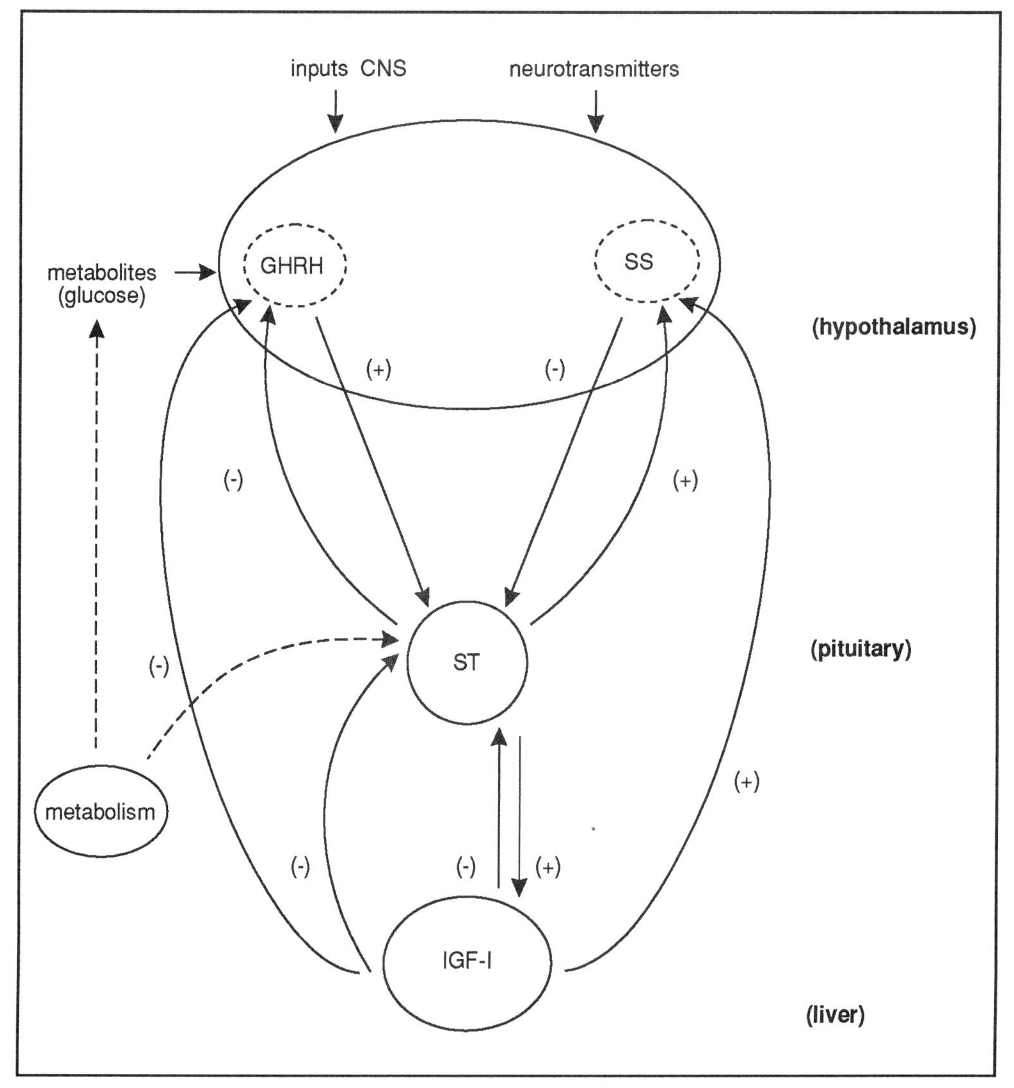

Figure 6.5 Diagrammatic representation of control of somatotropin secretion. Endocrine relationships are shown between the central nervous system (CNS; hypothalamus), pituitary somatotrophs and peripheral target tissues (liver, muscle). Somatotropin (ST) is regulated by stimulatory (growth hormone releasing hormone, GHRH) and inhibitory (somatostatin, SS) peptides from the hypothalamus, and secondarily by a number of peripheral feedback systems (eg by IGF-I = insulin-like growth factor).

6.2.1 Secretion pattern of somatotropin

St secretion pattern

The rate of secretion of ST in animals is not constant. In a number of mammalian species ST is released into the bloodstream in distinct episodes (peaks) throughout the day. In ruminants, the peaks of ST secretion occur apparently at random with periodicities which vary from animal to animal (Figure 6.6). It is tempting to ascribe a physiological meaning to these episodic pattern of secretion, notably the frequency of the secretory peaks, the amplitude of the peaks and the baseline ST concentration (an average of all observations less those which are part of a peak).

There are differences between male and female ST secretory patterns for some species. Also the stage of development plays a role. Ram lambs have higher frequencies and amplitudes of secretory peaks and baseline ST levels than ewe lambs or castrates; the same applies for cattle.

The episodic pattern of ST-release may be brought about by episodic release of GHRH and/or SS controlled by the central nervous system. However, there is a disagreement as to which of the two neuropeptides is ultimately responsible for the pattern of ST-release. A number of studies have suggested that SS withdrawal sets the timing of the episodic burst of ST-release while GHRH sets the magnitude.

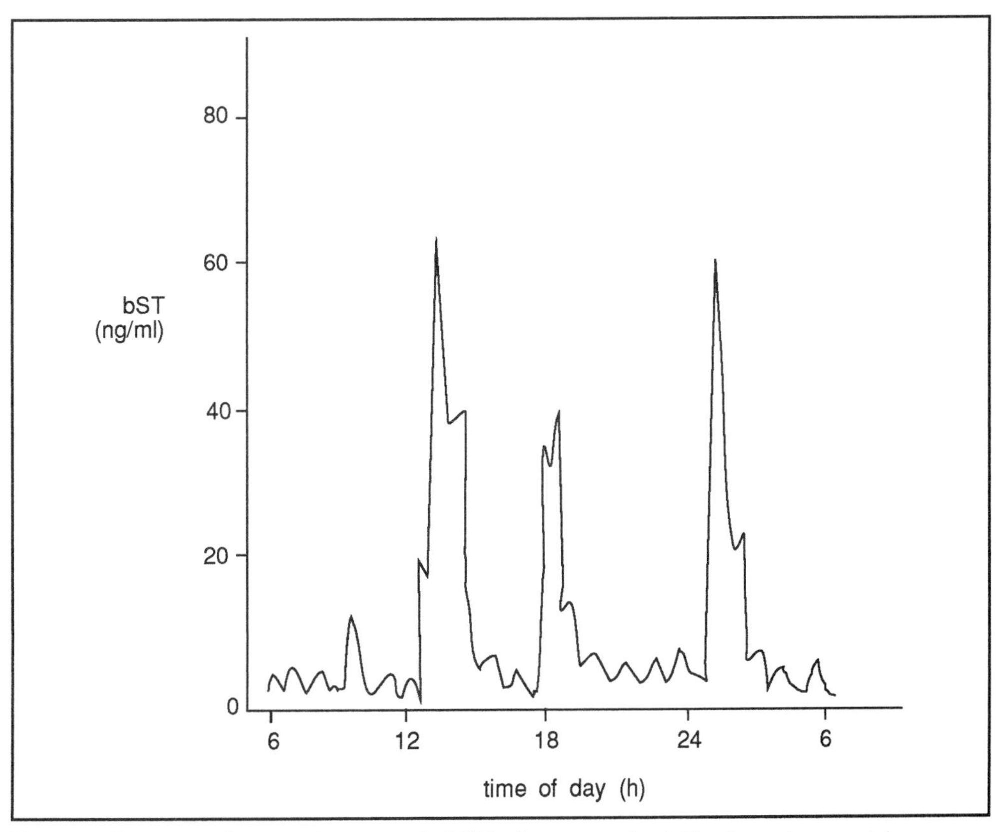

Figure 6.6 Representative secretion pattern for bST in the young animal. Blood samples were taken frequently over a period of 24 hours.

6.2.2 Insulin-like growth factors

During the last two decades it became clear that not all of the growth promoting actions ascribed to ST are direct actions of the hormone.

somatomedins or insulin-like growth factors

From massive quantities of human plasma, growth factors could be isolated and purified that were named somatomedins or insulin-like growth factors (IGF). In particular these polypeptide factors were found to mediate the ST-stimulation of linear growth *in vivo* by their mitogenic action on cartilage.

IGF-I and -II

The amino acid sequences of two peptides, IGF-I and -II, were determined and it appeared that they bear a striking structural similarity to the pro-insulin molecule. The primary sequences of human, bovine and porcine IGF-I are identical. In fact, on some tissues the IGFs act in an insulin-like manner and generally IGF-II has more insulin-like properties than IGF-I.

The main production site of the IGFs is in the liver; their biosynthesis, especially of IGF-I, is under the control of ST. Neither the liver nor any other organ however, were found to be enriched storage sites for the somatomedins. The IGFs are secreted as they are produced.

These 67-70 amino acids long polypeptides circulate in the bloodstream bound to specific carrier proteins, which are also produced in the liver partially under control of ST. About 99% of the circulating IGF-I is found complexed to its carrier proteins.

Hence GHRH, ST and IGF-I form part of the somatotropic control axis along which longitudinal bone growth and muscle growth are stimulated. However, the picture became more complicated when it was shown that many fetal and postnatal tissues and organs were able to produce IGF-I activity themselves. This may imply that growth and metabolic processes are locally regulated by the hormonal action of IGF-I in a so-called autocrine or paracrine way (autocrine or paracrine are terms used to describe a hormone-like activity which only acts over a short distance). It is largely unknown whether this is still under the influence of ST.

6.3 Endocrinology of somatotropin

ST has many effects on metabolism which are either directly or indirectly mediated by, for example, IGF-I. In fact the hormone appears to be able to modify many aspects of metabolism.

Broadly, two basic effects may be considered. The first is concerned with productive processes like growth of bone and muscle tissue and milk production, probably partly but not exclusively mediated by IGF-I. The second effect is on nutrient supply which is probably mediated by ST itself.

Π As you read through Section 6.3.1, it might be helpful to construct yourself a table summarising the information. Thus you could use the following headings.

action on	effect	direct or indirect action	metabolic basis for action
lipid metabolism	generally decreased adiposity		
	mobilisation increased	indirect	increases sensitivity of adipose cells to catecholamines

6.3.1 Effects on metabolism and cell division

Lipid metabolism and adipose tissue

Generally ST decreases adiposity which may be effected by changes in both lipid synthesis (mainly triacylglycerols) and mobilisation. As to lipid mobilisation, there is an indirect stimulating effect of ST on the ability of adipocytes (fat cells) to respond to lipolytic signals like the catecholamines (for example noradrenaline). This response is possibly enhanced by an increase of number of β-adrenergic receptors - binding sites for catecholamines in adipose tissue. A direct lipolytic effect of the ST molecule seems less probable. There is also an indirect effect of ST on lipid synthesis. ST antagonises the ability of the hormone insulin to maintain or increase lipogenesis. As a result glucose oxidation and fatty acid synthesis in the adipocyte are diminished.

Carbohydrate metabolism - liver

ST acts as a counter-regulatory hormone to insulin in carbohydrate metabolism, so inducing hyperglycemia. It decreases the ability of insulin to increase glucose utilisation by the tissues. ST must affect in some way the glucose transport system of the cell or its insulin receptor. In the liver, the processes of glycogen synthesis and glycolysis are not affected. On the other hand, liver glucose output is enhanced and there are indications that the liver gluconeogenesis is enhanced under influence of ST.

Protein and amino acid metabolism - skeletal muscle

It seems that ST has no effect on protein proteolysis or amino acid catabolism. ST does have an indirect effect on protein synthesis in muscle. This is brought about by a stimulation of amino acid uptake and incorporation into skeletal muscles. Probably the interaction of IGF-I with its muscle receptor is responsible for these processes.

Cellularity

ST has positive effects on cell proliferation (division) in a number of tissues and organs. These activities are largely mediated by IGF-I. Thus longitudinal bone growth is enhanced via a stimulated proliferation of cartilage cells. Also muscle cell proliferation (of so-called satellite cells) is under influence of IGF-I.

Π Can you offer any explanation(s) how of all these multiple actions of ST described above might find their basis in just one protein molecule?

There is no conclusive answer, however different regions of the molecule might provide for different biological effects. In this case these regions must be specifically recognised by heterogenous sets of ST receptors. Also after release of ST, proteolytic processing of the molecule may take place in the bloodstream or at the receptor site. This would generate smaller molecules each having a specific, restricted biological action with affinities for specific receptors.

6.3.2 Integrated processes

Physiological processes such as growth, gestation and lactation are very complex systems in which among other things nutrient fluxes are directed in a coordinated way. Hormones and their receptors play a prominent part in this coordination. However, it is clear that no single hormone can be responsible for any one of these life processes. Body growth for instance is controlled by a number of anabolic hormones - ST, IGFs, insulin, thyroid hormones, sex steroids - all working in concert. For example, in addition to being dependent on ST, the release of IGFs by the liver seems to require adequate levels of insulin. Similarly, the ability of IGFs to stimulate growth at the tissue level may be dependent on thyroid hormones. Thus a balance of hormones is required to stimulate growth maximally.

homeorhesis

In 1980 Bauman and Currie adapted an existing term 'homeorhesis' to describe the partitioning of nutrients which occurs during pregnancy and lactation. They defined homeorhesis as 'the orchestrated changes for the priorities of a physiological state ie coordination of metabolism in various tissues to support a physiological state'.

homeostasis

Homeorhesis is contrasted to homeostasis - maintenance of physiological equilibrium. Homeostatic control may be thought of as a regulation on a short-term basis while homeorhesis is more a long-term regulation occurring on a more gradual basis. Changes in body metabolism from pregnancy to lactation serve as an example of homeorhesis.

So, somatotropin may be seen as a hormone of homeorhetic control. The relatively high levels of ST in the early lactation of ruminants when animals are in negative energy balance helps in furnishing extra energy substrates and precursors for milkfat by an enhanced responsivity of lipid stores towards lipolytic signals, and also lipogenesis is suppressed. During later stages of lactation lipogenesis recovers when animals return to positive energy balance.

diabetogenic effect

The impairment of insulin action by ST (diabetogenic effect) facilitates the preferential utilisation of glucose by the mammary gland, for in lactating animals the mammary tissue is insensitive to insulin. Since very little glucose enters the ruminant's bloodstream, the animal must be very economical with its glucose supply. Both the lipolytic and diabetogenic effects of ST are means of diminishing glucose oxidation by muscle and other tissues.

SAQ 6.2

Select the appropriate words from the options given.

1) The change from childhood to adulthood requires homeorhetic/homostatic control.

2) ST increases/decreases the mobilisation of fats from fat cells.

3) ST antagonises/stimulates the ability of insulin to increase lipogenesis.

4) ST induces hypoglycaemia/hyperglycaemia.

5) It is thought that ST depresses/enhances gluconeogenesis in liver.

6) ST is thought to stimulate/inhibit the proliferation of cartilage cells.

6.4 Practical applications - history

hypophysectomy

Surgical removal of the pituitary and consequently the production site of a number of hormones is called hypophysectomy. Hypophysectomised laboratory animals provide a model for study of the influence of each, single pituitary hormone. The effects of pituitary removal are observed, then operated animals are treated with combinations of the pituitary hormones with or without the one actually under study.

Write down what effects on growth in rats would be expected after hypophysectomy followed by treatment with ST?

Your answer should have been that retardation of longitudinal- and weight growth are observed in young hypophysectomised rats. Growth can be restored by treating the animals for a time with ST given as a series of injections alone or in combination with other hormones.

6.4.1 Application in humans

In an analogous way, growth retardation caused by dysfunction of the hypothalamo-pituitary hormones in the prepubertal period of Man can be treated. Short stature of hypothalamo-pituitary origin is caused in most instances by ST deficiency.

Until recently treatment was with ST extracted and purified from human pituitaries collected at autopsy. ST has to be repeatedly injected over a period on account of its short biological half-life (10-20 minutes). During treatment, in young patients with a dwarfism syndrome, a catch-up growth response is seen, coupled to a prompt reduction in body fat and increases in muscle and bone mass.

6.4.2 Applications in farm animals

galactopoietic effect

Reports of the practical application of ST in the form of crude pituitary extracts in farm animals date from as early as the 1930s. Asimov and Krouze (1937) in the USSR were the first to demonstrate the galactopoietic (increased milk production) effect of crude anterior pituitary extracts in cows. They conducted a large scale trial with dairy cows receiving a single injection of anterior pituitary extract. After 12 days 170 treated cows had produced together more than 1900 litres of milk above the usual production.

In the 1940s this observation was confirmed by British researchers with a series of experiments. Also at this time bovine-ST was isolated in a highly purified state from bovine pituitaries and with this the galactopoietic effect of bST on the dairy cow could be definitely ascertained. It was found also that in goats, hypophysectomised during lactation, milk yield decreased drastically. It appeared that, of the complex of hormones needed to restore lactation, ST was of crucial importance.

It had been established therefore that in intact lactating ruminants injections of purified bST (exogenous ST) enhanced milk yield. This galactopoietic effect of ST was demonstrated during the ensuing 40 years in numerous short term experiments in dairy cattle and other ruminants.

influence on growth

The classical work of Evans and Simpson (1931) showed an increase in growth after chronically treating intact rats with crude preparations of somatotropin once the initial growth phase of the animal was over. Furthermore it was found that increase of body weight of rats was composed of protein accretion and reduction in fat deposition. So ST seemed capable of regulating both growth rate and the partition of nutrients between protein synthesis and fat deposition.

Π Unlike rats, for most species the effects of exogenous ST on length growth are limited to the initial phase of growth, up to puberty. Can you suggest why this is so? (You will need to know something about bone growth to be able to answer this).

This is closely connected with the influence of ST/IGF-I on the cartilaginous growing zone near the extremities of the long bones. From these growth plates bone tissue is being formed until the growth zones are replaced by bone and are closed. In rats this process is less strictly fixed during development.

Due to shortage of sufficient amounts of pituitary-ST, growth trials with farm animals were done on a small scale only. It was demonstrated that growth in hypophysectomised lambs could be restored by daily injections of purified bST.

Experiments with exogenous ST on intact growing meat animals gave mixed success. Positive effects of ST were an improved nitrogen-retention (indicating better protein production), an increase in the ratio of lean meat (muscle) to body fat, and an increase in the efficiency of food conversion. But any other effect on total body weight gain was often offset by a substantial reduction in adipose tissue. It also became clear that the nutritional state of the animal and the feeding regime (protein level and energy value) greatly influenced the extent of the effect of exogenous ST.

6.5 Alternative approaches to manipulation of production

Besides administration of somatotropin to the animal, other strategies have been devised with the same ultimate aim: enhancement of the mean blood level of somatotropin and/or IGF-I to increase efficiency of production.

SAQ 6.3

Can you suggest an alternative treatment that would increase blood ST? Reference to Section 6.2 and Figure 6.5 should provide a clue.

As well as GHRH treatment, reducing the concentration of somatostatin (SS) reaching the pituitary has been another approach to stimulate the release of ST. This was done by actively immunising young growing animals against somatostatin. The SS was made immunogenic by coupling it chemically to human serum globulin. The animal's immune system recognises this conjugate as non-self and produces antibodies which include anti-SS antibodies. These will complex circulating SS and consequently SS in the portal system of the pituitary is reduced. Though in some cases such immunoneutralisation of SS was accompanied by growth promotion (lambs, calves), this effect was not invariably seen and in the case of growth enhancement, there was not always a rise of mean levels of ST. The mechanism of this growth promotion is not clear especially as somatostatin is also produced as a regulator peptide in other parts of the body (gut, pancreas).

immuno-neutralisation of SS

Little is known about the growth-promoting potential of exogenous IGF-I in farm animals. Until recently, the somatomedins have only been available in very small quantities so experimentation with IGF-I and -II has been with small laboratory animals only. Administration of human-IGF-I to hypophysectomised rats or dwarf mice was found to increase body length and weight, in a manner similiar to that found with ST, although the responses were not as effective and organ weights were differentially stimulated. In all cases the free IGF-I (not bound to its carrier proteins) was given by injections. Hence the question arises whether the carrier proteins are prerequisite for an effective biological action.

However if, for example, IGF-I increases growth rate in pigs, such treatment will not be comparable to the effects seen with exogenous pST. This is because the metabolic effects of pST in adipose tissue will not be mimicked by IGF-I.

Recombinant-DNA derived somatomedins are now becoming available and much more research can be foreseen. In the next section we will examine this technology.

Note that human, porcine and bovine IGF-I appear to have identical primary structures.

6.6 Recombinant-DNA bovine somatotropin

The only site of ST production in the cow is in its pituitary. In the somatotrophs (cells which produce ST) the bST-gene is transcribed as mRNA and the bST-mRNA is translated into protein. So bST-mRNA contains all the information which is also found in the protein's primary structure.

6.6.1 Structure of bST-mRNA

The obvious organ for isolation of bST-mRNA is the bovine pituitary because the gene of interest is expressed here. Figure 6.7 outlines the process used to determine the structure of bST-mRNA. Use it to follow the description given below.

mRNA

After isolation of total RNA, the mRNA fraction (polyadenylated at the 3′ end) was purified by chromatography on oligo (dT) cellulose to which the poly A-messengers stick.

cDNA

Next all mRNA molecular species - including the bST-mRNA - were converted enzymatically (with reverse transcriptase) into double stranded copy DNAs. The bST-mRNA molecule could be expected to be within the range of approximately 1000 base pairs calculated from the number of amino acids in bST.

cDNA bank

In order to have sufficient amounts of cDNA available for sequence analysis, a size fraction of 600-1000 base pairs of ds cDNA was cloned into the plasmid pBR322-DNA which was amplified in *E. coli* as host organism. Colonies from this 'cDNA-bank' were grown in microtitre wells and aliquots of the cultures were assayed for their bST-cDNA sequence.

probe

On account of the evolutionary conservation of the ST-sequence, an existing radiolabelled probe of rat-ST cDNA could be used. Hybridisation of the unknown cDNA to this probe revealed the presence of bST sequences.

From the cDNA sequence of the excised plasmid-DNA the bST-mRNA structure was deduced.

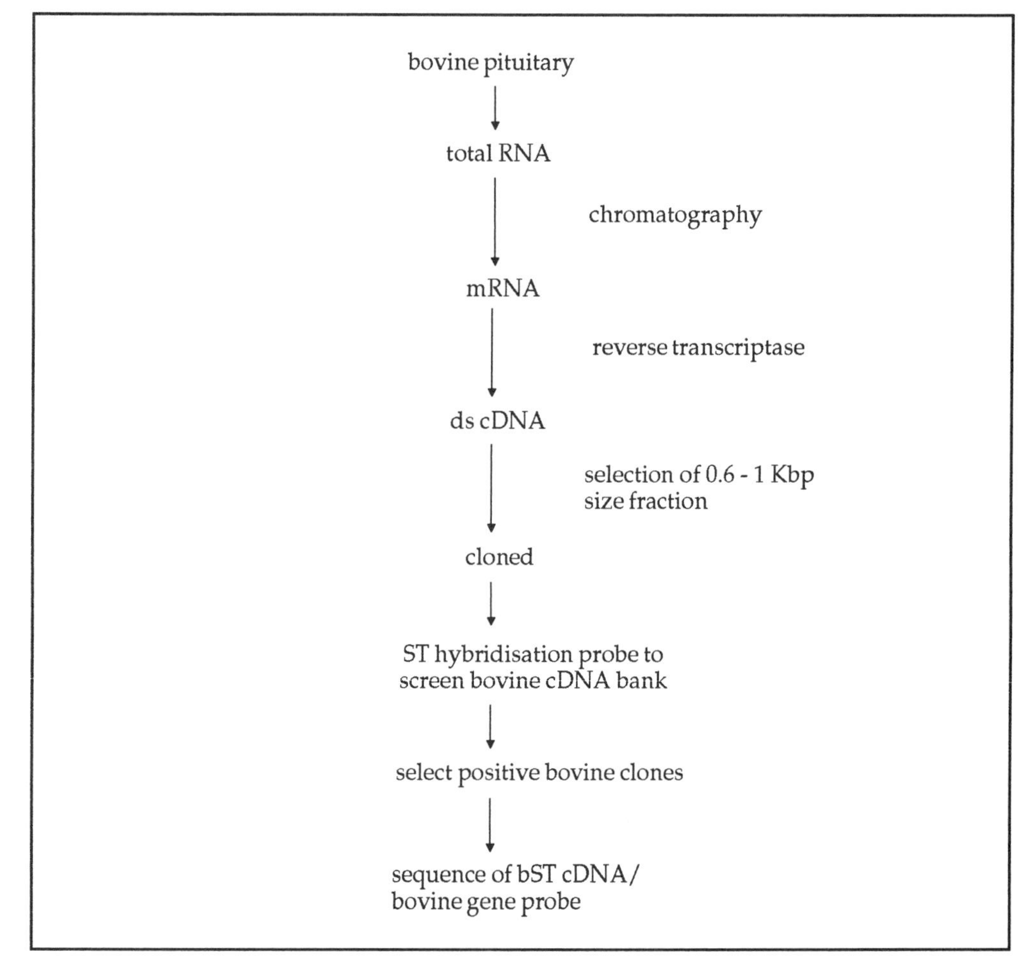

Figure 6.7 Flow diagram of the technology used to generate bovine gene probes for ST and/or starting material for elucidation of the structure of bST mRNA.

The messenger structure has a length of about 800 nucleotides and appears to stimulate the synthesis of a protein of 217 amino acids and about 23 kD molecular weight, which is higher than that of the mature bST structure of 190 amino acids.

Π Can you explain this apparent anomaly in the size of the synthesised peptide?

signal peptide

The 217 amino acid precursor form contains a so-called signal peptide of 27 amino acid residues covalently linked to the N-terminus of the mature hormone. The precursor is translated on polysomes attached to the rough endoplasmatic reticulum (RER). The nucleotide and protein sequences are given in Figure 6.8. Note the non - translated sequence of nucleotides at each end of the mRNA.

```
                         5' end AGGAUCCCAGGACCCAGUUCACCAGACGACT-
                                         ——— signal peptide ———
                                         Met  Met  Ala  Ala  Gly  Pro  Arg
-CAGGGUCCUGUGGACAGCUCACCAGCU AUG AUG GCU GCA GGC CCC CGG -
                                         -27                          -21
 ________________________________________
 Thr  Ser  Leu  Leu  Leu  Ala  Phe  Ala  Leu  Leu
 ACC UCC CUG CUC CUG GCU UUC GCC CUG CUC UGC
 -20                                      -10
 ______________________________________
 Leu  Pro  Trp  Thr  Gln  Val  Val  Gly  Ala  Phe  Pro  Ala  Met
 CUG CCC UGG ACU CAG GUG GUG GGC GCC UUC CCA GCC AUG ...
                                 -1   +1

                                         Cys  Ala  Phe
                                     ... UGU GCC UUC UAG UUGCCA
                                                190

-GCCAAUCUTUUGUUUGCCCCUCCCCCGUGCCUUCCUUGACCCUGGAAGGUGCC-
-ACUCCCACUGUCCUUUCCUAAUAAAAUGAGGAAAUUGCAUCGC poly (A) 3' end
```

Figure 6.8 Nucleotide sequences around the 5' and 3' ends of bST-mRNA. The protein sequence of the signal peptide and the terminal parts of bST are also shown. Note that the structure presented here for the mature bST is the 190 amino acid variant.

The signal peptide, with a high content of hydrophobic amino acid residues, helps the precursor molecule to cross the membrane of the RER of the somatotroph. After passage across the membrane the signal peptide is split off enzymatically and the mature protein is released into the intracisternal space. From here it migrates to the secretory granules of the somatotroph.

In an analogous way, the structures of messengers for ST of human, porcine and other species were elucidated. In this way the hitherto not completely unravelled protein sequence of pST could be established. Besides some homology in the mature protein structure, there was a high degree of homology for the signal peptides of the ST of different species.

6.6.2 Genomic organisation of bovine somatotropin

After isolation and characterisation of the cDNA coding for bST, the chromosomal gene for bST was also studied in detail.

Genomic DNA from the bovine pituitary was isolated and digested with restriction enzymes. DNA fragments were cloned into a vector system and amplified in *E. coli*. Next this amplified genomic library was screened for the presence of bST sequences with the help of a labelled hybridisation probe. Positive clones were selected and the genomic DNA of interest excised from the vector plasmid and sequenced.

Comparison of the nucleotide sequence of bST-mRNA and the corresponding chromosomal DNA indicated that the bST gene has a discontinuous structure (Figure 6.9). Five different parts in the genomic DNA (exons) also present in the bST-mRNA are interrupted by four intervening sequences (introns) not present in the bST-mRNA. The entire gene is approximately 1800 base pairs long.

As is characteristic of eukaryotic transcription, the longer primary transcription product synthesised in the nucleus of the somatotroph, after splicing, yields the bST-mRNA.

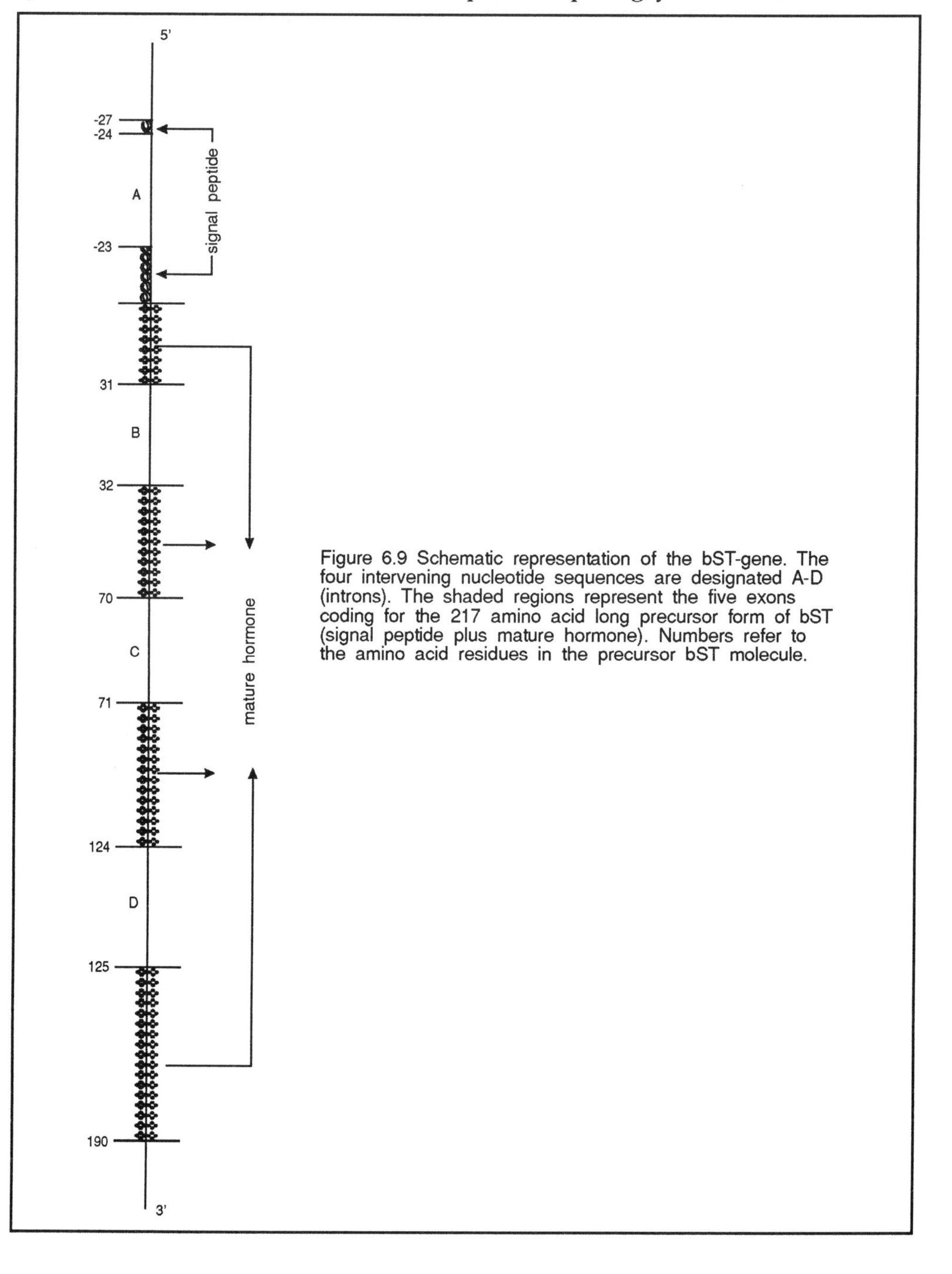

Figure 6.9 Schematic representation of the bST-gene. The four intervening nucleotide sequences are designated A-D (introns). The shaded regions represent the five exons coding for the 217 amino acid long precursor form of bST (signal peptide plus mature hormone). Numbers refer to the amino acid residues in the precursor bST molecule.

6.6.3 Production of ST via recombinant-DNA techniques

For the bioproduction of the eukaryotic somatotropins in bacteria certain requirements are necessary for the successful expression of the corresponding genes. There are several important differences in gene structure and function between prokaryotes and eukaryotes and this causes some problems in using eukaryotic genes in prokaryotic (bacterial) systems.

Π Can you write down some problems that might arise due to these differences?

Key differences between gene structure and function in prokaryotes and eukaryotes mean that:

1) the eukaryotic bST gene is not suitable for expression in bacteria due to its discontinuous structure. Prokaryotes do not have the enzymes necessary for mRNA splicing;

2) expression of bST-DNA in bacteria is not simply achieved. Bacterial transcription and translation signals are needed. For this purpose a prokaryotic promoter upstream from the gene has to be inserted;

3) the bST-cDNA, containing the continuous bST-gene is coding for the precursor molecule and bacteria are not able to remove enzymatically the signal peptide yielding the mature bST molecule.

In order to overcome the difficulties mentioned under 1) to 3), bST-ds cDNA was chosen as the basis for a DNA construct for final bacterial expression (Figure 6.10). As this DNA contains the unwanted information for the signal peptide, the corresponding sequences must be removed. This was done by substituting a small fragment in the ds DNA containing the coding sequence for the signal peptide plus part of the N termini of bST by a stretch of synthetic DNA.

First the *Pvu II* fragment carrying the coding sequence for bST amino acids 23-186 was isolated. This large natural fragment was ligated to a synthetic short fragment (labelled as Met 1-22 in Figure 6.10) containing sequences for the first 22 amino acids plus an N-terminal methionine (ATG).

As this ligated fragment - with codons for amino acids 1 to 186 - did not yet contain the complete information for bST, substitution with a fragment coding for the C-terminal part of bST was the next step.

The ligation product was therefore cleaved with the endonucleases *Eco*RI and *Pst* I. This fragment was inserted into pBR322 linearised by the same enzymes to yield a plasmid vector with the 1 to 90 coding sequence.

To complete the hybrid gene construction, a *Pst I* fragment from the cloned bST-ds cDNA, containing the codons for bST residues 91-190 (plus the 3′ untranslated portion of bST-ds cDNA) was inserted into the single *pST I* site of the above mentioned plasmid vector (containing the codons for residues 1-90).

As a prokaryotic promoter is needed for this otherwise complete structural bST-gene, the *E. coli trp* promoter was used. This has been used before to get a high expression of the human somatotropin in *E. coli*. For that reason the bST coding region was installed downstream from the *trp* promoter in a new vector plasmid and cloned.

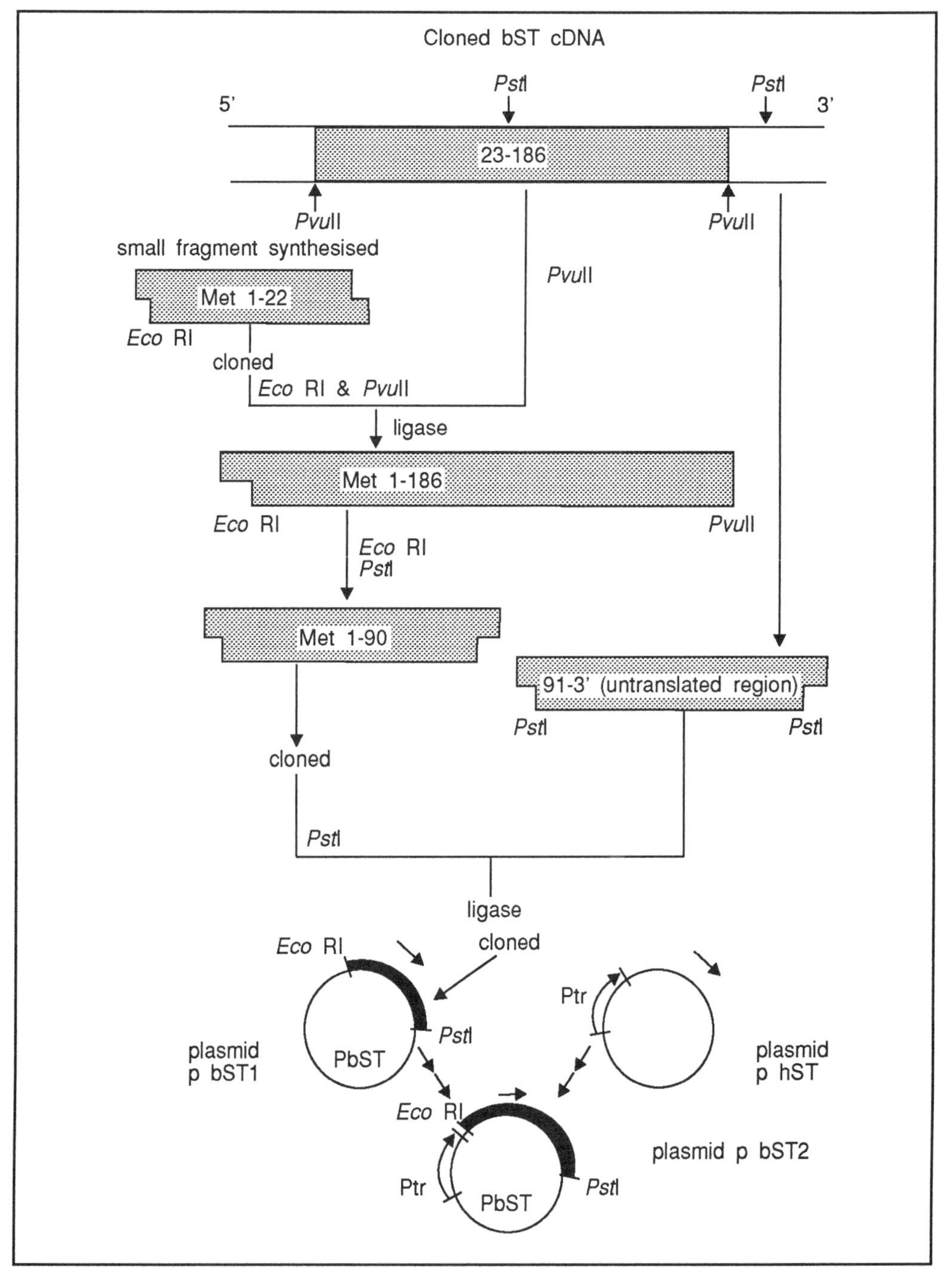

Figure 6.10 Simplified diagrammatic representation of the construction of a bacterial expression vector (p bST2) for bST. Plasmid bST1 contains a synthetic-natural hybrid gene coding for bST. The tryptophan promoter (Ptr) was taken from an existing plasmid (p hST) used for the expression of mature hST. Arrows show 5'-3' direction of coding sequences (see text for further details).

Bacterial cultures transformed with these plasmids were grown under conditions known to induce *trp*-promoter -directed expression. Yields as high as 1.5g of bST per litre of bacterial culture could be obtained. bST was found within the *E. coli* cells in an aggregated form packed in distinct cytoplasmatic granules. It had to be solubilised with a denaturing agent (disrupting the secondary and tertiary structures) and it became evident that the hormone was expressed in reduced form (cysteines instead of disulphide bridges). It was possible to refold the molecule by air oxidation and then the natural form with two disulphide bridges could be obtained in a high yield. The hormone had to be scrupulously purified to remove bacterial and culture medium contaminations. This product was not completely identical to the hormone as secreted by the pituitary. It had an extra N-terminal methionine.

In a similar way to that described above, human-ST had already been obtained, and also porcine ST (pST) could be bio-engineered. Other workers prepared recombinant-DNA bST identical to the naturally occurring variant with N-terminal sequence (Met)-Ala-Phe-Pro... or rec-DNA bST with a small extra N-terminal extension of eight amino acid residues.

Moreover, recombinant human somatotropin has now been successfully prepared without the N-terminal methionine. This is of particular importance on account of a possible antigenic effect of the methionine-hST during treatment of human patients.

SAQ 6.4

If the gene manipulations described above for the construction of plasmid p bST2 (Figure 6.10) had been carried out using the promoter for β-galactosidase instead of the tryptophan promoter, which of the following would be true or false.

1) The bST product would not have been produced in a reduced state.
2) The bST product would have been produced in a solubilised form.
3) bST production would have been greater if the host was cultivated in a glucose medium.

6.7 Application of recombinantly derived bST in cattle and pST in pigs

In the dairy cow absorbed nutrients are partitioned between maintenance of body tissues, milk production, growth of body tissues (which might occasionally be mobilised) and eventually growth of the foetus. bST partitions nutrients towards milk production and away from body tissue storage. In the growing animal, for example the fattening pig, pST benefits the rates of bone and muscle growth at the expense of growth of fatty tissues.

The biotechnological production of STs with recombinant DNA techniques made it possible to study the application of STs in large-scale experiments. In the 1980s, hundreds of trials were performed in dairy cattle and growing pigs. These major experimental programmes were funded by the pharmaceutical companies who developed the recombinant somatotropins. Their production by chemical synthesis of (cheaper) steroid anabolics to improve beef production hampered the use of bST in large-scale experiments on beef production. The use of pST in growing pigs was more successful than in lactating sows. In the short lactation period of the sow a very large part of the fatty tissues is usually mobilised for the lactation process and this might have prevented a favourable effect of pST treatment in the sow on the weight gain of her litter.

Firstly, results of bST-application in cattle will be reviewed. Secondly, the effectiveness of pST-application in growing pigs will be outlined.

6.7.1 Effects of bST-application on milk production

In small scale trials daily treatment with pituitary bST for some days increased milk yield (+ 3.3 kg/day) and fat percentage in the milk (+ 0.25%) and decreased protein in the milk (- 0.11%). In a few trials over some months these figures were respectively: + 3.4 kg/day, - 0.07% and 0.00%. In these experiments extracted pituitary bST was compared to the recombinantly derived bST. Based on daily injection the latter gave better results: + 5.7 kg milk/-day, + 0.03% milk fat percentage and + 0.04% milk protein percentage. Impurities in the pituitary - derived ST might be responsible for the better response to the recombinant ST.

Using simple organic substances the pharmaceutical companies developed slow-release formulations to avoid daily injections. In dose - response studies the optimum dose of bST was found to be 20-40 mg bST per day. Subcutaneous injections of bST in a slow-release formulation (500 mg/14 days or 640 mg/28 days) yielded a response of 4.0 kg milk/day, - 0.04% fat percentage and + 0.01% protein percentage. On average the 500 mg dose injected once a fortnight increased milk yield by 4.4 kg/day and the 640 mg dose injected once every 28 days gave a response of 3.1 kg milk/day.

For a number of reasons it is recommended not to start bST application in dairy cows for the first three months after calving. Just after calving, milk production increases faster than feed intake. Usually a dairy cow has a negative energy balance in the first 6-8 weeks after calving, in which body tissues are mobilised for milk production. In this period bST treatment has a lower efficacy. In general treatment before day 90 of lactation (cows are rebred between 60-90 days of lactation) gave lower pregnancy rates, which is generally believed to be caused by the negative energy balance. This is initiated through the rise in milk yield after the first bST-injection. Six to eight weeks after the first treatment feed intake increases to meet the requirements for the higher milk production.

In general bST had no effect on digestibility of the ration and no effects on the maintenance requirements per kg metabolic body weight. Metabolic body weight is body weight raised to the power of 0.75 ($W^{0.75}$) and is better related to basal metabolism. Bovine ST treatment had a large effect on body fat mobilisation. In order to meet feed requirements for maintenance and the increased milk production, lactating treated cows could store less body fat than control cows fed the same ration (25-70 kg body lipids at the end of the treatment period). Despite these effects, levels of metabolites in the plasma of treated cows (for example 3-hydroxy-butyric acid and non-esterified fatty acids) never indicated metabolic disorders.

If cows to be treated are not fed properly and the energy content of the diet is low then treatment with bST will yield only a slight increase in production. In pasture conditions (grass feeding and a low concentrates level) bST treatment was about 50% less effective than at winter rations with a high level of concentrates. A high energy diet with an adequate protein supply is required to deliver glucogenic and aminogenic precursors to enhance milk production after bST treatment while body fat tissues are a prerequisite for mobilisation during the period of negative energy balance after the first injections. Responses in milk production to bST treatment show a large variation within cows, between the treatment periods in a lactation (reproducibility = 20%) and within cows between lactations (= 50%). Systematic factors, which might influence the response such as breed of the cow (dairy, dairy/beef), parity (number of calvings), or number of lactations treated in succession, do not interact with bST-treatment. The low reproducibility of the response in milk production indicates the impossibility of detecting good and bad responders to bST treatment after a single injection.

So far, the slow-release formulations used to inject bST subcutaneously once every 14 or 28 days do not give a steady response in milk production during the treatment period as can be seen in Figure 6.11. bST levels in blood during the treatment period revealed a similar curve as for milk yield per day. Such a cyclic pattern of production should be avoided, because of disrupting a proper milk recording system and also giving rise to a cyclic pattern in fat and protein percentages in the milk. The latter may have a slight effect on dairy processes.

Π On Figure 6.11, mark the times of bST injection on the horizontal (time) axis and note their coincidence with the intermittent rises in milk production in the treated cows.

A review of numerous trials with bST treatment indicates that no adverse effects of treatment on the health, metabolism or offspring of treated cows were found. Fertility is not affected when cows are refertilised before the onset of bST treatment.

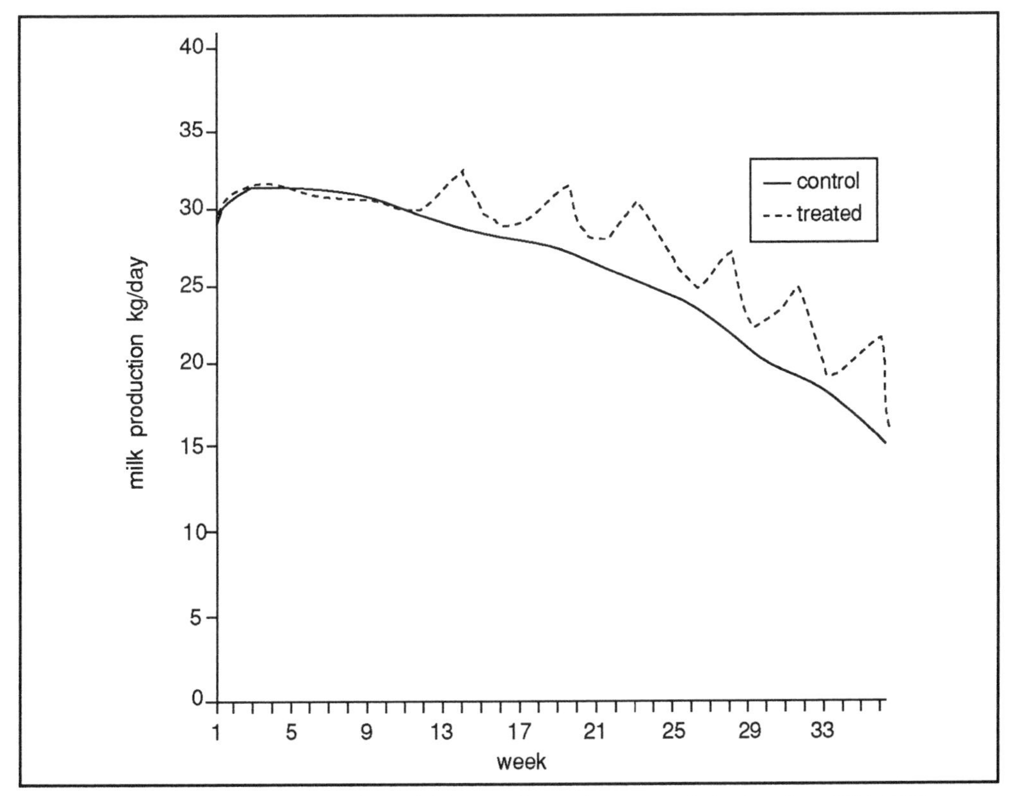

Figure 6.11 Milk production of control- and bST-treated cows. Treatment (injection of rec-DNA bST in a sustained-delivery vehicle) was started at week 13 of lactation and was repeated over 5 consecutive 28-day periods. The levels of exogenous somatotropin were 640 mg and zero (untreated controls). Note rec-DNA bST = bST produced via recombinant DNA technology.

6.7.2 The mechanism of action of bST on milk production of dairy ruminants

While much is known of the effect of exogenous bST on lactation, the mechanism of action of bST has not been fully established.

Up to now, three physiological components of the enhanced milk response can be identified.

1) Exogenous bST causes an altered nutrient utilisation and mobilisation in the non-mammary tissues which spares essential nutrients for milk synthesis (Section 6.3.2). This is brought about by direct effects of bST on the liver and adipose tissue function.

2) In response to bST treatment substantial increases in mammary blood flow have been observed in goats and cows. However, it is not clear whether this increased flow and consequently a greater supply of substrates to the udder, is a cause or a consequence of the increase in milk secretion. The latter seems the more likely possibility.

3) bST alters mammary gland function itself by changing its secretory capacity. There is strong evidence that bST does not have a direct effect on the udder. Receptors for bST were not found in bovine mammary tissue and direct infusion of bST into the artery to the udder of lactating ewes did not induce an enhancement of milk response.

SAQ 6.5

How might this observed response of bST treatment on secretory capacity be explained (refer to Section 6.2.2)? (Hint - think about IGFs)

6.7.3 Application of bST in heifers reared for milk production

mammogenesis

bST has a stimulating effect on mammogenesis (mammary gland development) before the onset of puberty. Daily injections of 20 mg bST between 8-12 months of life resulted in an increase of udder secretory cells of up to 46%. It is expected that this will result in a larger milk production capacity after calving. However, this effect has yet to be proven.

6.7.4 Effects of pST-application in growing pigs

As stated before, only a few trials with bST on growing cattle have been performed. On the average, bST-treatment increased daily gain (12%), had no effect on feed intake, favoured feed conversion (9%), increased meat content of the carcass (5%) and decreased its fat content considerably (15%).

The ability of somatotropin to alter glucose, fat and protein destinations is of specific interest for meat production in growing pigs. The quality of pig meat is based worldwide on the percentage of lean meat in the carcass and there is a continuing pressure to reduce fat content. Therefore the ability of somatotropin to increase lean tissue deposition at the cost of fat deposition is of great importance to the pig industry. A pST dose level of 30-60 μg/kg body weight/day results in up to a 5% reduction in feed intake, a 20% increase in growth rate, a 20% reduction in the quantity of feed required per kg live weight gain, a 10% increase in lean content of the carcass and a reduction of 30% in the fat content of the carcass. The latter is of special value for the fatty American pig breeds, which can be slaughtered with a much better carcass composition at a fixed age/weight after pST treatment. With the leaner European pigs the optimum carcass composition can be achieved at a higher weight/age ratio by pST application.

Practical application of pST is hindered by the absence of a slow-release formulation for the pig, which necessitates daily injections. The induced change in protein/fat deposition calls for a higher protein content of feedstock for pigs to be treated with pST.

6.8 Effects of ST-treatment on milk- and carcass quality

For this purpose quality will be defined as the value of the animal products for human nutrition and their safety for human consumption (absence of residues).

6.8.1 Value of milk from bST treated cows for dairy processing

milk quality

The fat, protein and lactose content of milk do not alter when bST-treated cows have a positive energy- and protein balance (consume more feed than is required for maintenance and milk production). A negative protein balance, mainly caused by inadequate rations, will reduce the protein content of the milk. Negative energy balances, which appear just after calving and after the first bST injections, result in an increase of long chain fatty acids in milk originating from mobilisation of body fat and therefore in an increase of the fat content of the milk. To a small extent, this phenomenon influences the 'spreadiness' of butter. In cheese processing no differences in cheese traits between milk of bST-treated and control cows have been found.

6.8.2 Value of meat of ST - treated animals

meat quality

So far, no negative effects of ST treatment on meat quality parameters have been reported. The tenderness of meat might decrease, when due to ST treatment of lean cattle and pig breeds, less fat is deposited in the carcass. For tenderness meat requires a certain amount intramuscular fat.

6.8.3 Safety of the consumer of milk and meat from ST-treated animals

safety for the consumer

bST treatment of dairy cows increases ST concentration in plasma considerably: 5-9 fold. However, no increase can be measured in the very low bST level in milk. Shortly after calving, the IGF-I level in milk is high: 150 ng ml^{-1}. As a result of bST treatment during lactation it increases by 3 - 12 ng ml^{-1} (in plasma from 100 to 400 ng ml^{-1}).

SAQ 6.6

Why should even increased levels of bST and IGF-I pose no or very low risks to the human consumer?

6.9 Profitability of application of bST to dairy cows

In many countries where dairy production is an important economic enterprise, milk production per farm is limited to avoid storage of large surpluses. In this situation a higher milk production per cow results in less cows per farm and so cannot increase gross income per farm. Therefore, in many countries bST-application can only be used to save on the costs of milk production. Profitability of bST cannot be estimated precisely, because its price and the costs of application are not known (for the moment).

Model calculations show that profitability depends mainly on the number of cows per unit of pasture and on the opportunity to develop alternative non-dairy enterprises at the farm. A relatively low number of dairy cows per hectare and/or the absence of viable alternatives to dairy production will not permit profitable application of bST.

Under some conditions, bST-application to dairy cows can increase profitability of farms despite quotas on milk production. In countries with seasonal differences in milk price it can assist in producing more milk in the months with higher milk prices. At the end of the year it can allow the milk quota to be reached with a higher priced milk when, for example poor weather has limited milk production.

So, despite the higher efficiency of treated cows (more milk per unit of feed) resulting in lower costs per kg milk, bST application currently does not generally result in a higher profitability for the farm.

6.10 Socio-economic effects

From the first fantastic results (+ 41% milk/day) in the USA and assumptions that all dairy cows would be treated with bST, it was concluded that bST application would have a drastic effect on herd size and in numbers of dairy farms. Using slow-release formulations and assuming a more realistic figure of 40% of the dairy cows to be treated and an increase of 10% milk yield per cow, leads to more likely results. Under these conditions cattle numbers in the EC would reduce by 1% annually due to the application of bST.

Selective use of bST within a farm aiming to create cows having milk production levels acceptable for breeding on a large scale by artificial insemination, would decrease genetic progress for milk production in a population by up to 7 per cent of the original level. It is impossible to detect bST-treated cows from the milk record data of herds because of the large variation in milk production between cows in a herd. Registration of bST-treatment will show a large variation in regimes. Besides, registration cannot easily be checked. Plasma bST levels of a treated cow can be as high as found during 'peaking' in an untreated cow (Figure 6.6). Selective use of bST within the farms supports the need to concentrate potential bull dams at a limited number of strictly controlled farms. New reproduction techniques in cattle will also stimulate this trend.

6.11 Registration as a veterinary drug

EC registration authorities, who examine new biotechnological products, consider bST and pST as veterinary drugs. Therefore they must be tested for effectiveness and for safety to both the animal and the consumer of animal products. Within the EC a company may request registration of a biotechnological product in one of the member states. This state sends the application in combination with all available trial reports to the European Commission for Veterinary Drugs. All EC members participate in this Commission, which has 90 days to assess the effectiveness and safety of the drug. Additional questions may lead to a further 90-day delay. The advice of this Commission to the member states is not binding, though it must be fully considered when individual states decide on approval within 30 days. In this procedure biotechnological developments are not considered with respect to their political or social aspects.

6.12 Ethical issues

We concluded the previous chapter by raising the question of the acceptability (or otherwise) of genetically manipulating animals. In much the same way we can ask the question whether it is acceptable to manipulate the growth and (metabolism) behaviour of animals by using 'drugs'. Although this does not have the same long term and self-perpetuating consequences that can be ascribed to genetically manipulating animals, it certainly impinges on the area of animal rights. Whether or not these practices are acceptable depends, of course, on individual beliefs and upon the perceived costs and benefits of these processes.

Summary and objectives

This chapter has dealt with the molecular biology, physiology and applications of growth hormone or somatotropin (ST) and related factors. The effects and use of somatotropin in human medicine and animal production have been reviewed briefly. A detailed scheme for the manufacture of recombinant somatotropin was given and the effects of its use on milk and meat production in livestock animals was described. Lastly, the potential impact of the widespread use of somatotropin on agricultural practices was considered. On completion of your study of this chapter you should be able to:

- describe in broad terms the molecular biology of somatotropin and the mechanisms that regulate its secretion and mediate its function, including the role of growth factors;
- understand the applications of somatotropin treatment in human medicine and animal models and their relevance to animal production;
- describe the basic techniques used to produce recombinant bovine somatotropin;
- describe the major effects of somatotropin treatment on milk and meat production in livestock animals;
- make judgements concerning the risk to humans arising from somatotropin treatment.

Vaccines and diagnostics

Vaccines and diagnostics

7.1 Introduction

Vaccines are indispensable in the economic production of food animals. In the control of highly contagious animal diseases (eg foot-and-mouth disease and swine fever), vaccines have played and still play a crucial role. Vaccines have been mainly developed against viral diseases, but the increasing resistance of bacteria to antibiotics and the problem of antibiotic residues in animal products have re-activated interest in vaccines against bacterial diseases. The rise of contemporary biotechnology, more than a decade ago, has revolutionised vaccinological research and has led to large investments in research on the development of modern vaccines. At present, however, only a handful of 'biotech' vaccines have reached the marketplace. The original wave of optimism has waned and has given way to more realistic views and time schedules on the development and marketing of such vaccines. In spite of this somewhat disappointing progress, biotechnology has created tools for research on the genetic structure and immunogenicity of microbes, and on the molecular pathogenesis and immunology of infectious diseases. The knowledge gained in this field is providing the basis for modern approaches to vaccine development.

7.2 A brief history of vaccinology

Edward Jenner and Louis Pasteur were the founders of vaccinology. Jenner is reported to have carried out the first widely publicised attempt to protect humans against smallpox by deliberately inoculating them with material derived from a pustular lesion of a human affected by cowpox. The inoculated humans developed an immunity against smallpox. Jenner called the material vaccine, from the Latin word 'vacca' for 'cow', and the inoculation process 'vaccination'. He published his spectacular findings in 1798, in an article entitled: 'An inquiry into the causes and effects of the Variolae Vaccinae: a disease discovered in some of the western counties of England, particularly Gloucestershire, and known by the name of cowpox'. The Jennerian approach to prevent smallpox proved to be so effective, that as an eventual result, and due to a strategic vaccination campaign, the World Health Organisation could officially announce the global eradication of smallpox on 8 May, 1980.

vaccine

vaccination

Do you think Jenner was the first to attempt to prevent infectious diseases by a 'vaccination' procedure?

No. Approximately 1000 A.D., in India, it became the practice to deliberately inoculate, either intradermally (into the skin) or by nasal insufflation, pustular material from human pox lesions. It resulted in an infection that was usually less severe than a natural infection. From India, this practice spread to China, West-Asia, Africa and, in the beginning of the 18th Century, to Europe.

attenuation

The next major advance in the field of vaccinology after Jenner was Pasteur's work, in the 1870s, on attenuation of the bacterium (*Pasteurella multocida*) that causes fowl cholera. Pasteur observed that a culture of this organism that was accidently left exposed to air over a holiday period, no longer induced disease in chickens, but provided immunity against a challenge exposure to a fresh culture of *Pasteurella multocida*. He perceived that the principle of immunity was the same as Jenner's, and to honour Jenner, he proposed henceforth use of the term 'vaccine' for all products intended to prevent infectious diseases. Pasteur established the principle of attenuation, as he recognised that organisms could be rendered avirulent by various treatments, but still were able to evoke immunity. We will discuss the issue of attenuation and avirulence in later sections. This concept of attenuation also underlied the successful development of the rabies vaccine that Pasteur first administered to human beings in 1885.

Since Pasteur, the principles for production of live vaccines have essentially remained unchanged. In the 1930s, Theiler developed the 17D vaccine strain of yellow fever virus, which is still in use, by attenuating a virus strain through more than 200 passages in mouse embryonic tissue culture and subsequently in chicken embryo tissues.

The golden age of vaccine development began however at the end of the 1940s, when antibiotics became available. It was then possible to culture living cells on a large scale. Frenkel, working in Holland with foot-and-mouth disease virus, was the first to grow a virus on a scale large enough for mass vaccination to be undertaken. Since that time, other viruses have been grown in sufficient quantities to produce vaccines.

In the final years of the 20th century, a new wave of vaccine development is about to occur, mainly as the result of biotechnology (Table 7.1). Although most vaccines used in Man and animals are still produced by conventional means, several 'biotech' vaccines have already reached the marketplace (see Section 7.6).

1973	First cloning of a gene
1974	First expression of a gene cloned in a bacterium
1975	First hybridoma cells for production of monoclonal antibodies
1978	First recombinant DNA product available (somatostatin)
1981	First diagnostic kit based on monoclonal antibiotics
1982	First recombinant DNA vaccine licenced for use in livestock (colibacillosis in pigs and cattle)
1986	First recombinant DNA vaccine licensed for use in man (hepatitis B)

Table 7.1 Milestones in biotechnology.

7.3 Immune response to vaccination

The objective of vaccination is to provide the best possible protection for an animal against an infectious disease. Generally, a natural infection with a micro-organism affords a strong immunity against re-infection.

∏ What ideal features should vaccination possess compared with natural infection?

A vaccination should simulate a natural infection without producing disease, but induce the same level of protective immunity, and particularly stimulate immunological memory.

7.3.1 Structure and function of the immune system

We have assumed that you are familiar with the immune system. The description below is written to remind you of the important features of the system. If you have no immunological experience then you may need to refer to a text on the immune system. We recommend the BIOTOL text 'Cellular Interactions and Immunology'.

The structural elements of the immune system are the central and peripheral lymphoid organs. The central lymphoid organs consist of bone marrow, thymus, and in avian species the bursa of Fabricius (which is associated with the gut). From these organs the lymphoid cells arise. Lymph nodes, spleen, and mucosa-associated lymphoid organs, such as Peyer's patches, constitute the peripheral lymphoid organs, and these are the sites of primary immune responses.

Bone marrow is the source of stem cells, from which the precursor lymphocytes arise (Figure 7.1). These are destined to become either T lymphocytes or B lymphocytes or a so called 'null-cell' subset. The pre-T-lymphocytes enter the thymus gland, where they undergo maturation and differentiation, and then leave the thymus. These lymphocytes, upon contact with antigen, will enter the peripheral lymphoid organs where the final stages of differentiation occur.

Maturation and differentiation are reasonably complete when B lymphocytes leave the bone marrow, except in avian species where differentiation occurs in the bursa of Fabricius. The mature B cells enter the circulation and peripheral lymphoid organs. The lymph nodes are the site of most antigen-driven immune responses.

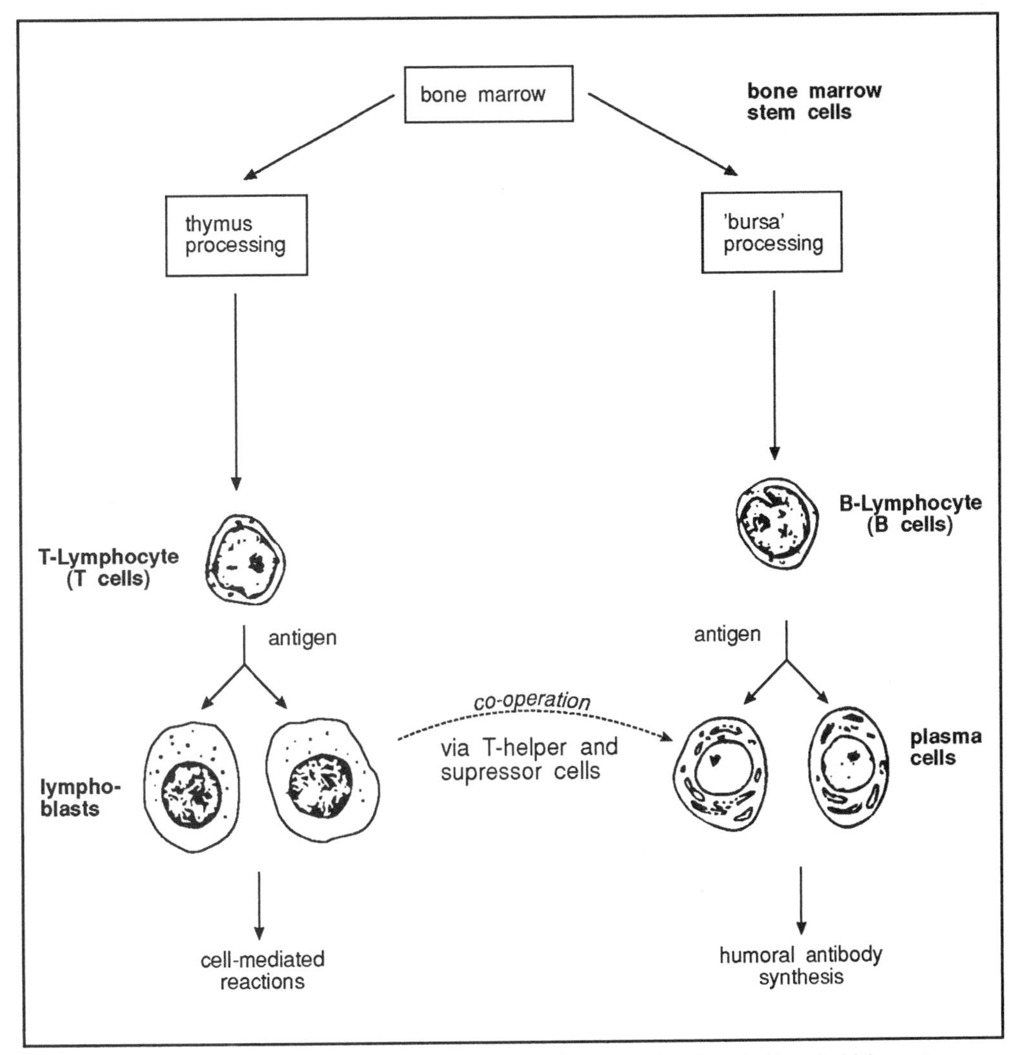

Figure 7.1 Processing of bone marrow cells by thymus and gut-associated central lymphoid tissue to become immunocompetent T- and B-lymphocytes respectively. Proliferation and transformation to cells of the lymphoblast and plasma cell series occurs on antigenic stimulation.

The cooperative interaction between antigen-presenting cells, T lymphocytes and B lymphocytes in response to antigen results in the activation and proliferation of T and B lymphocytes, differentiation of B lymphocytes into plasma cells that secrete antigen-specific immunoglobulins (humoral immunity), and in the generation of T-helper and T-suppressor lymphocytes and specific T-effector lymphocytes involved in cell-mediated immunity. In addition, T- and B-memory lymphocytes are produced. They recirculate through peripheral lymph nodes, spleen and blood for many months or years. An encounter between the memory cells and the antigen results in a rapid memory response (or secondary response) consisting of cellular proliferation, differentiation and antibody synthesis that is quantitatively greater than a primary response. In Figure 7.2 antibody responses in control (non-vaccinated) and vaccinated pigs are compared following challenges by the virus causing Aujeszky's disease (see Section 7.3.3). Note that after vaccination, exposure to the virus causes a rapid and substantial production of antibodies (ie secondary response). If the animal has not been

humoral immunity

cell-mediated immunity

memory or secondary response

previously exposed (ie the controls), then a slow production of antibodies occurs (ie primary response). The rapid secondary response means that antibodies are rapidly produced to counteract infection. Immunological memory therefore provides the basis for protective immunity following a second exposure to the antigen.

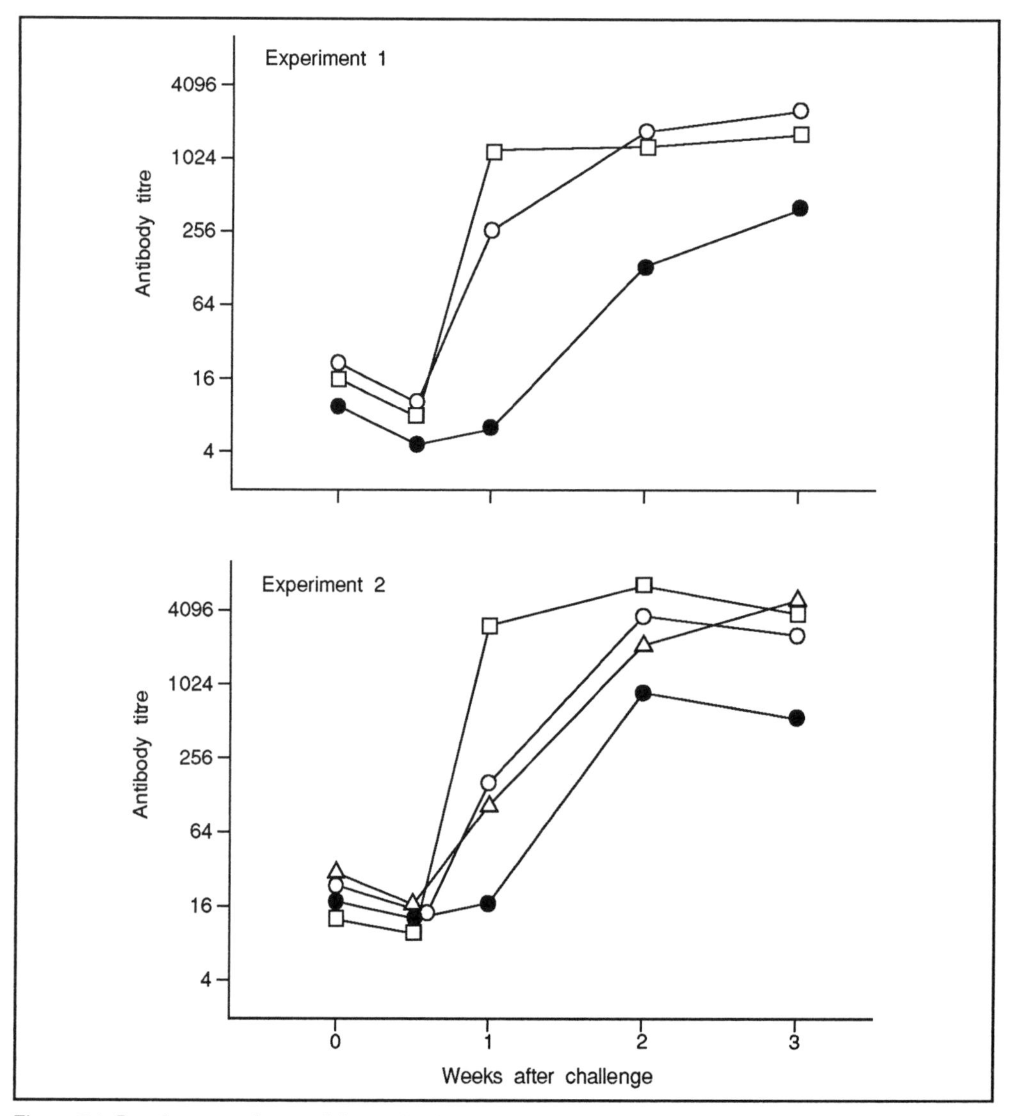

Figure 7.2 Development of neutralising antibody response to Aujeszky's disease virus in pigs (mean geometric titre), (n = 6 to 8) after challenge exposure. ● —— ● controls, ❑ —— ❑ intranasal vaccination (vaccine A), O —— O vaccine B, Δ —— Δ vaccine C. (Data derived from van Oirshot (1987), Research in Veterinary Science 42 p12-16) see text for further details.

systemic immunity

Figure 7.2 deals with the antibody responses measured in serum, that is systemic immunity. Contrasted to this is mucosal immunity which operates relatively independently from the systemic immune system. Antigen is presented to the mucosal immune system by modified epithelial cells known as M cells. These M cells are capable of sampling antigens from the intestinal lumen and transporting them across the cell to the underlying lymphoid tissue of the Peyer's patches. Upon antigen-induced

mucosal immunity

proliferation both B and T lymphocytes enter the mesenteric lymph nodes for additional maturation. Thereafter, they migrate into the lamina propria of the intestine and respiratory tract, and to secretory organs such as salivary glands and mammary glands. Most of the B lymphocytes will differentiate into IgA-producing plasma cells. Note that IgA is a particular class of antibodies (immunoglobulins). They are characterised by being the immunoglobulin that is predominantly found in mucosal secretions. The mucosal immunity is expressed at most, if not all, mucosae of the animal. To induce a vigorous mucosal immune response a local or mucosal administration of micro-organisms is required.

SAQ 7.1

What disease characteristics are likely determine whether vaccination should be aimed at inducing a systemic or a mucosal immunity?

7.3.2 Maternal antibodies and vaccination

maternal antibodies

Antibodies are not transmitted from mother to young during pregnancy in ruminants and pigs. Newborn animals receive antibodies from their mother through ingesting colostrum (early 'milk') shortly after birth. Because the intestine is permeable to immunoglobulins during the first 24-36 hours after birth, these maternal antibodies reach the blood stream of the young. These passively acquired maternal antibodies, which usually have a biological half life of two to three weeks, protect the young animals against disease due to infection with various pathogens. They are indispensable for the survival of young animals. However, one major disadvantage of the presence of maternal antibodies is their interference with vaccination. Maternal antibodies strongly inhibit the development of antibodies (Figure 7.3) and of active immunity after vaccination and consequently the animals are less well protected than those that are free of maternal antibodies at vaccination. The degree of inhibition is inversely related to the level of maternal antibodies the animals have at vaccination.

Π What implications does this inhibition have for animals reared for food?

Food animals usually live for only a short period, for example chickens six weeks and pigs six months, so vaccination of these animals is often hampered by the presence of maternal antibodies for a significant period of their life. In the case of Aujeszky's disease an intranasal vaccination has been developed that overcomes the inhibition by maternal antibodies rather better than does an intramuscular vaccination.

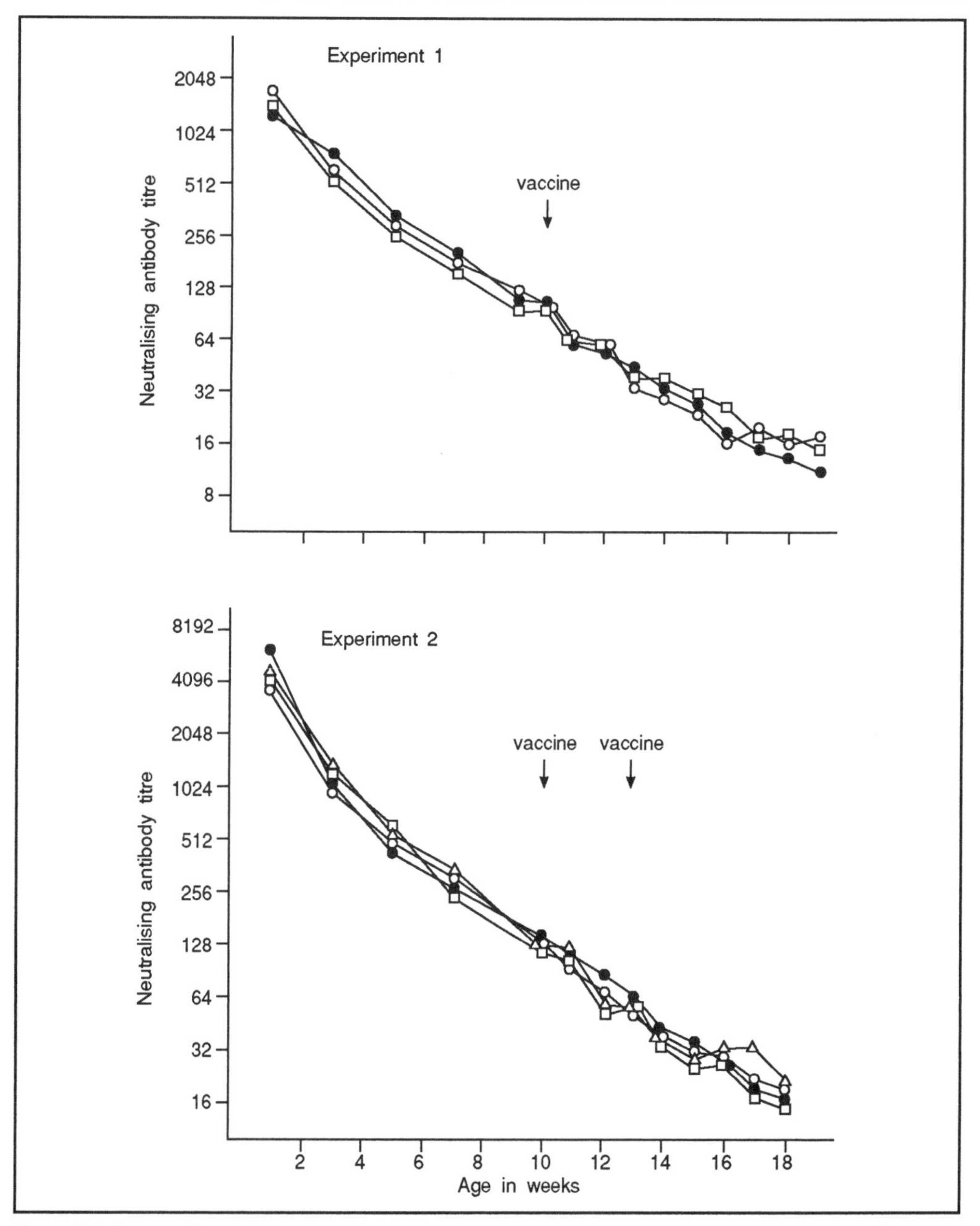

Figure 7.3 Course of mean (n = 6 to 8) neutralising antibody titres before and after vaccination against Aujeszky's disease of young pigs. Note the general decline in maternal antibody levels and the inhibited response to vaccination. ● —— ● controls, ❑ —— ❑ intranasal vaccination (vaccine A), ❍ —— ❍ vaccine B, Δ —— Δ vaccine C. (Data from van Oirschot (1987), Research in Veterinary Science 42 p12-16).

7.4 Live and killed conventional vaccines

Π Make a list of criteria that we might apply to an ideal vaccine.

We would include in our list the following features.

An ideal vaccine simulates the natural infection to a high degree, without causing disease or any other adverse effect. A single vaccination should induce a strong and long-lasting protective immunity. A vaccine should be stable, easily administered, adaptable to mass vaccination and cheap. Furthermore, the antibody response after vaccination should be distinguishable from that due to natural infection, so that vaccination and eradication may proceed simultaneously. Let us explain this a little more fully.

Let us say a vaccine stimulates the production of antibodies with specificity A, B, C, D whereas the natural disease stimulates the production of antibodies with specificity A,C,D,E. In principle we can use the presence or absence of antibodies of E specificity to identify animals that have been infected (or are infected) by the natural disease agent. These animals can then be eradicated.

Π Write down what you think the most crucial task is for a vaccine.

A vaccine must induce immunological memory, which is responsible for a rapid and overwhelming secondary immune response upon infection.

Now we know what we are trying to achieve in producing a vaccine, we can examine the strategies for vaccine production.

Almost all of the vaccines currently in use in the veterinary field are produced by conventional means. There are two broad categories:

live vaccines

- the attenuated, or modified-live vaccines (also referred to as live vaccines);

killed vaccines

- the inactivated or killed vaccines. This category also comprises the toxoid vaccines (see Section 7.4.2).

The features of each category are described below.

7.4.1 Live vaccines

virulence

immunogenicity

For use in domestic animals live vaccines are only available against viral diseases, not against bacterial diseases. The live vaccine usually contains an attenuated virus able to replicate in the host. The aim of attenuation, first recognised and utilised by Pasteur, is to weaken or eliminate the disease - causing properties, (the virulence) of the virus, without affecting its immunity - inducing properties, its immunogenicity. Because virulence and immunogenicity are two closely linked genetically determined properties of a virus, attenuation will generally affect immunogenicity as well. The process of attenuation is often much the same as in Pasteur's time. It is essentially a matter of 'trial and error'. Viruses are given multiple passages usually in cell cultures or in animal species to which they are not naturally adapted. After a certain number of passages the virus is tested for vaccine efficacy and safety. When it is not sufficiently attenuated the cell-culture passages are continued, until a virus is obtained that no longer induces disease. It is clear that the vaccine producer can hardly influence this largely empirical process of attenuation and that its results are unpredictable.

SAQ 7.2

Attenuation can be considered, in some ways, as a primitive form of genetic engineering. Can you explain this view?

Using this empiracal approach a number of highly efficacious vaccines have been developed. For example, a vaccine strain against swine fever has been developed by more than 800 serial passages using defibrinated blood, or spleen or lymph nodes of inoculated rabbits. This so-called Chinese strain is highly efficacious and safe. A single vaccination induces a long-lasting immunity. Vaccinated pigs are fully protected against a challenge infection that kills all the unvaccinated control pigs. Furthermore, vaccinated pigs do not excrete or transmit the challenge virus to contact pigs, a feature crucial to eradication programmes. In the Netherlands, swine fever vaccine may only be applied in emergency vaccination programmes including a national eradication campaign. The vaccine has been proven to be indispensable for the successful eradication of swine fever.

7.4.2 Killed vaccines

adjuvant

Killed vaccines against viral and bacterial diseases are available for use in domestic animals. A killed vaccine contains a micro-organism that is rendered non-infectious by inactivation and therefore does not replicate in the host. To produce killed viral vaccines, the virus is grown in large quantities, and after a purification or concentration step, it is usually chemically inactivated. The inactivation process is aimed at destroying the infectivity of the agent while retaining its immunogenicity. To induce sufficient immunity, it is necessary to add an adjuvant (from the Latin word 'adjuvare' meaning 'to help', see also Section 7.4.4) to killed vaccines. The Netherlands' cattle population is still vaccinated annually with an inactivated vaccine against foot-and-mouth disease. This consistent vaccination programme has provided the basis for the elimination of foot-and-mouth disease virus in most Western European countries. Unfortunately, on the other hand, the production of the vaccine and its application have also caused a substantial number of disease outbreaks in Western Europe (Figure 7.4).

Π How do you think that the production of the vaccine and its application have caused a substantial number of outbreaks.

Vaccine production plants and incomplete formalin-inactivated vaccines have been shown to be sources of virus dissemination. This finding is one rationale for attempts to design a foot-and-mouth disease vaccine by modern techniques that banishes the use of large quantities of virulent virus. The latter disadvantage of foot-and-mouth disease vaccine production also initiated the international discussion aimed at abolishing this vaccination process in the European Community in 1992.

toxoid vaccine

A toxoid vaccine is one in which a toxin is modified such that it is no longer toxic but will stimulate the production of antibodies which will neutralise the natural toxin.

An example of a toxoid vaccine is the tetanus vaccine. The bacterium *Clostridium tetani* is cultured, and the toxin produced is purified and inactivated and adsorbed onto an adjuvant. Hence the antibody raised neutralises the toxin rather than the micro-organism.

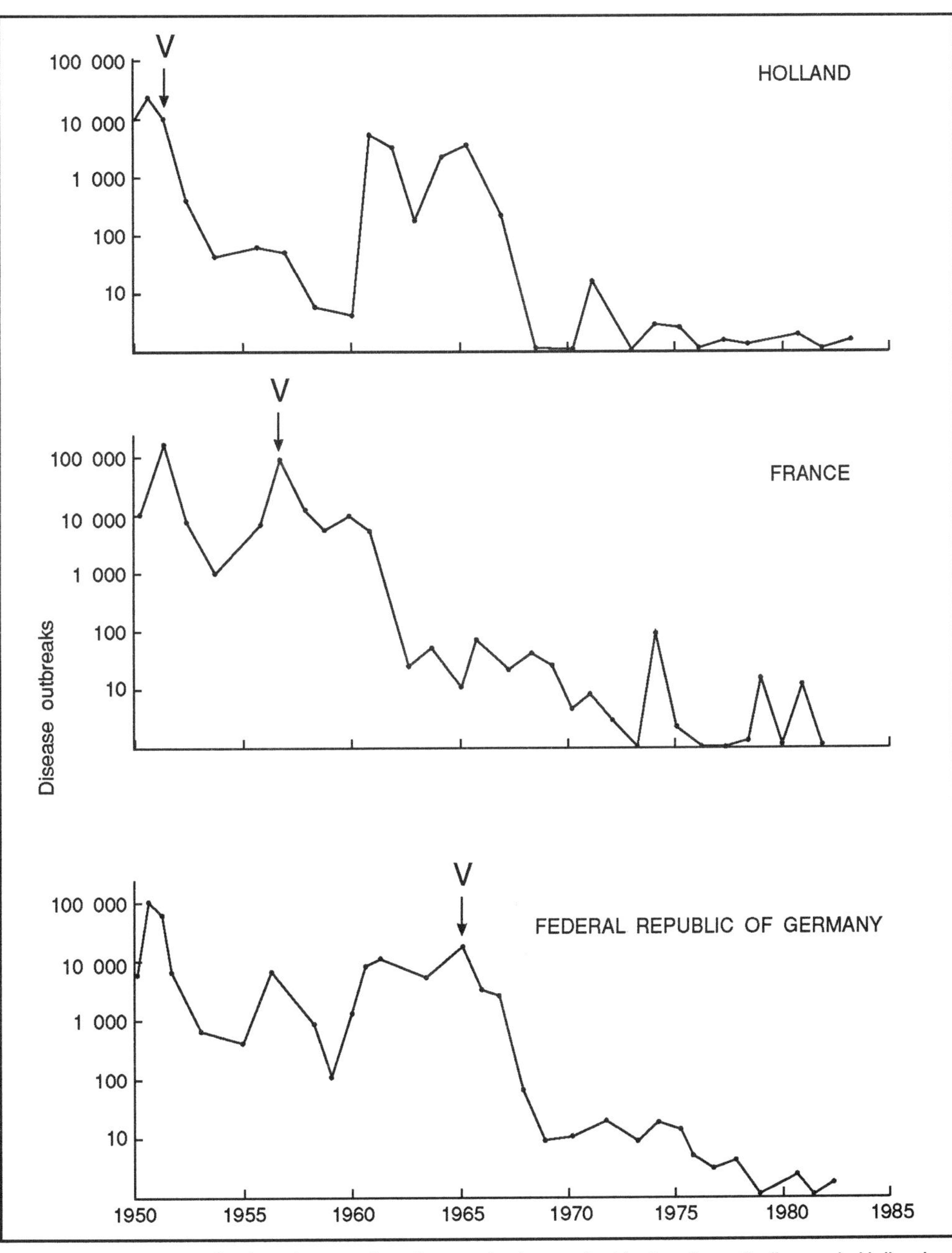

Figure 7.4 Effect of application of comprehensive vaccination against foot-and-mouth disease in Holland, France, and West Germany on the number of outbreaks of the disease. V, Start of mass vaccination. (Data from Brown (1989), Advances in Veterinary and Comparative Medicine 33 p173-178).

SAQ 7.3

Early in Section 7.4 features of an ideal vaccine are given. Do you think a live or killed vaccine will best meet the efficacy requirements of an ideal vaccine? Give your reasons.

7.4.3 Advantages and disadvantages of live and killed vaccines

In the previous SAQ, we gave a specific answer to the question of what makes an ideal vaccine. In fact the answer is not so clear-cut, it is a continuous matter of debate whether a live or killed vaccine represents the vaccine of choice. There is, however, no general answer to this question. It varies from one infectious disease to the other, and is dependent on the molecular biology of the micro-organism, the pathogenesis and immunology of the disease, and the prevalence of infection. Where the basic requirements for a vaccine such as induction of long-lasting protection and harmlessness are met by a killed vaccine, this product would be preferable. However, this is often not the case. When considering the advantages and disadvantages of live and killed vaccine efficacy and safety are the two main criteria.

Π Can you think of any other criteria for evaluating vaccines? (Make a list).

We have produced two. It is important that the vaccines batches can be made of the same quality. Thus, reproducibility is important. The cost of the vaccine doses is also important because, in the livestock industry, the decision to vaccinate or not is frequently based on economic factors, estimated by cost-benefit analysis.

Efficacy

It should be obvious that a live micro-organism, which replicates in the vaccinated host, better mimics a natural infection than a non-replicating micro-organism, thereby potentially mobilizing the different parts of the immune system. During its replication various antigens and non-structural proteins are produced. Several antigens are exposed on the surface of infected cells, which is necessary to induce memory cytotoxic T cells. Non-structural viral proteins, which are proteins that are not incorporated in the virus particle, can elicit a protective immunity. Examples are proteins which are involved in the maturation of a virus but do not themselves become incorporated into the viral particles. Immunity against such proteins would provide some protection. Another feature of a live vaccine is the possibility of delivering it onto mucous membranes, resulting in local replication and the induction of a mucosal immunity. This may be particularly advantageous for animals that are vaccinated in the presence of maternal antibodies. Often a single vaccination with a live vaccine gives rise to a solid and enduring protective immunity.

The above features do not apply to killed vaccines. It appears that the systemic humoral immune response is triggered after vaccination with a killed vaccine. Repeated vaccinations are required to induce a high level of protective immunity. The antibody titres after vaccination with a killed vaccine may be higher than after vaccination with a live vaccine. Consequently, offspring that ingest colostral antibodies from animals vaccinated with a killed vaccine may possess higher antibody titres than offspring from dams given a live vaccine and therefore may be passively protected better and longer.

Safety

Killed vaccines are safe with respect to residual virulence, vaccine persistence in host's tissues, reversion to virulence and transmission of vaccine to unvaccinated hosts. These advantages of killed vaccines correspond to the disadvantages of live vaccines: some may possess residual virulence, not only for the target host, but also for other animals. Some live vaccines have been shown to persist in a latent state in the vaccinated animal, for example bovine herpes virus type I vaccines. Although it is not often the case, some vaccines may revert to virulence and thus cause disease after vaccination. Reversion to virulence may primarily occur in viruses with a single stranded RNA genome, because they have a high mutation frequency. For many veterinary vaccines it is not known what is the underlying modification of the genome responsible for attenuation. A vaccine strain that carries one mutation on its genome may more readily revert to virulence after replication than a vaccine containing a strain that has multiple genetic alterations. In the poultry industry, where spray (aerosol) vaccinations are often applied, transmission of vaccine virus between hosts is considered an advantage, rather than a disadvantage.

contamination of live vaccines

Live vaccines always present the risk of contamination with unwanted organisms; for instance, outbreaks of pestivirus infection among pigs have been caused by contaminated Aujeszky's disease vaccines. Outbreaks of reticulo-endotheliosis in chickens in Japan and Australia have been traced to contaminated Marek's disease vaccines. Mycoplasma species may also contaminate veterinary vaccines. In 1986, it was shown that 7.5% of the commercial vaccine batches examined were contaminated with mycoplasmas.

The disadvantages of killed vaccines correspond to the advantages of live vaccines. The use of adjuvants can cause severe local and systemic reactions, while multiple dosing increases the risk for hypersensitivity reactions.

SAQ 7.4

Table 7.2 lists the main pros and cons of live and killed vaccines. Having read Section 7.4.3 assign each item on the list to the appropriate category of vaccine by marking '+' in the column to indicate a comparative advantage.

1) Efficacy	**Live**	**Killed**
a - broad immunity		
b - durable immunity		
c - mucosal administration		
d - number of vaccinations		
2) Safety		
a - residual virulence		
b - persistence		
c - reversion to virulence		
d - transmission		
e - recombination with wild-type virus		
f - contamination with extraneous agents		
g - local and systemic side-effects		
h - incomplete inactivation		
3) Cost		

Table 7.2 List of pros and cons of live and killed vaccines.

7.4.4 Adjuvants

To potentiate their efficacy, killed vaccines require the addition of adjuvants. An adjuvant can be defined as any substance that nonspecifically enhances the immune response to a given antigen. A good adjuvant should be safe and should produce an earlier, better, and more enduring immune response than the inactivated vaccine material alone. A large number of compounds have adjuvant activity, but only a few are actually incorporated into veterinary vaccines, among them are insoluble salts, such as aluminium hydroxide or aluminium phosphate, oil-in-water emulsions, water-in-oil emulsions, or Quil A. Quil A is purified from saponin, which is an extract of a South-American tree *Quillaja saponaria molina.*

The mechanisms underlying the potentiating activity of adjuvants are only partly elucidated. Adjuvants exert a depot effect: antigen is sequestered at the inoculation site and is slowly released into the systemic circulation, thus prolonging the lifespan for the antigen and its exposure to cells of the immune system. Adjuvants can also activate lymphocytes and influence lymphocyte regulation.

It is often emphasised that the local and occasional systemic reactions caused by adjuvants are unacceptable. Recently, some countries have prohibited the use of mineral oil emulsions in vaccines, because of their adverse effects on the living animals, and because their wider distribution may make parts of the carcass unsuitable for consumption. However, a certain level of an inflammatory response is required to induce an effective and durable protective immunity. It is a difficult task to develop adjuvants that stimulate a controlled and beneficial inflammatory process with minimal adverse effects.

7.5 Efficacy and safety of vaccines

As we have seen the two most important attributes of a vaccine are its efficacy and safety. In this section we will examine these a little more closely.

7.5.1 Efficacy

vaccine efficacy

Vaccine efficacy can be defined as the degree to which it induces a protective immunity in the target host. The induced protection has to be evaluated primarily in the target animal. Thus, for instance, vaccines against Aujeszky's disease are tested in pigs and vaccines against infectious laryngotracheitis in chickens. The vaccine efficacy can be evaluated according to various procedures, but basically it goes as follows. A number of animals is allotted to two equal groups. One group is vaccinated with the vaccine, the other group serves as unvaccinated control. After a certain period of time, both groups are inoculated, if possible, via the natural route of infection with a virulent micro-organism. After this challenge infection, the animals of both groups are monitored for development of disease signs or pathological lesions, for replication or excretion of the micro-organism, or for other biological characteristics that are intended to be prevented by vaccination.

Two examples are given:

1. Parvovirus infection of pigs (swine fever)

parvovirus infection

A parvovirus infection is only harmful to pregnant sows when their foetuses are exposed before the 60-70th day of pregnancy. Such infected foetuses die and, dependent on the gestational development, are either resorbed or mummified. Foetuses infected beyond the 70th day of pregnancy mount an immune response to parvovirus and survive the infection without experiencing any harm. Therefore, an efficacious vaccine must prevent the transmission of parvovirus from sow to foetuses during the first 9-10 weeks of pregnancy. The efficacy of vaccines is tested by vaccinating sows before breeding and leaving a group of sows unvaccinated. Six to seven weeks after breeding all pregnant sows are experimentally infected with a virulent virus. Several weeks later, they are killed and their foetuses, dead or alive, are examined for the presence of parvovirus. In the unvaccinated control sows foetuses must be infected, whereas in the sows vaccinated with an efficacious vaccine no infected foetuses may be present. Thus, a parvovirus vaccine can be considered efficacious if it prevents the transplacental transmission of a field strain of parvovirus.

Does the vaccine-induced immunity persist for one, two or three pregnancies? Although this question is of practical importance, such studies on long-term immunity are usually not examined in efficacy tests.

2. Aujeszky's disease in pigs

Aujeszky's disease

This is a widespread disease of pigs caused by a herpes virus. A standardised test for comparing the efficacy of vaccines against Aujeszky's disease has been developed. For this purpose, 10-week-old specified pathogen-free pigs, having no antibodies to Aujeszky's disease virus, are divided into two groups of eight pigs. One group is vaccinated, the other is left unvaccinated. The groups are housed separately. Three months after vaccination all pigs are intranasally challenged with 10^5 plaque-forming units (infectious doses) of the virulent NIA-3 strain. After challenge, the pigs are monitored for mortality and clinical symptoms, including fever. They are weighed and the number of days of growth arrest is determined. In addition, the excretion of challenge virus per ml of saliva is measured. Comparing several vaccines according to the above procedure allows the most efficacious vaccine(s) to be detected. An Aujeszky's disease virus vaccine is efficacious when it reduces the periods of growth arrest and virus excretion by more than 50%, by comparison with unvaccinated pigs (Table 7.3 and Figure 7.5).

Vaccine	Mean SN-titre at challenge	No. dead/ No. tested	Challenge results Mean days of		
			Growth arrest	Fever*	Virus shedding
A	70	0/8	8	3	5
A 2 x	700	0/8	2.5	1	5
Controls	-	3/8	16	5	9

Table 7.3 Summary of results of standardised tests to evaluate efficacy of vaccines against Aujeszky's disease. *Fever = body temperature above 40°C. Growth arrest in days.

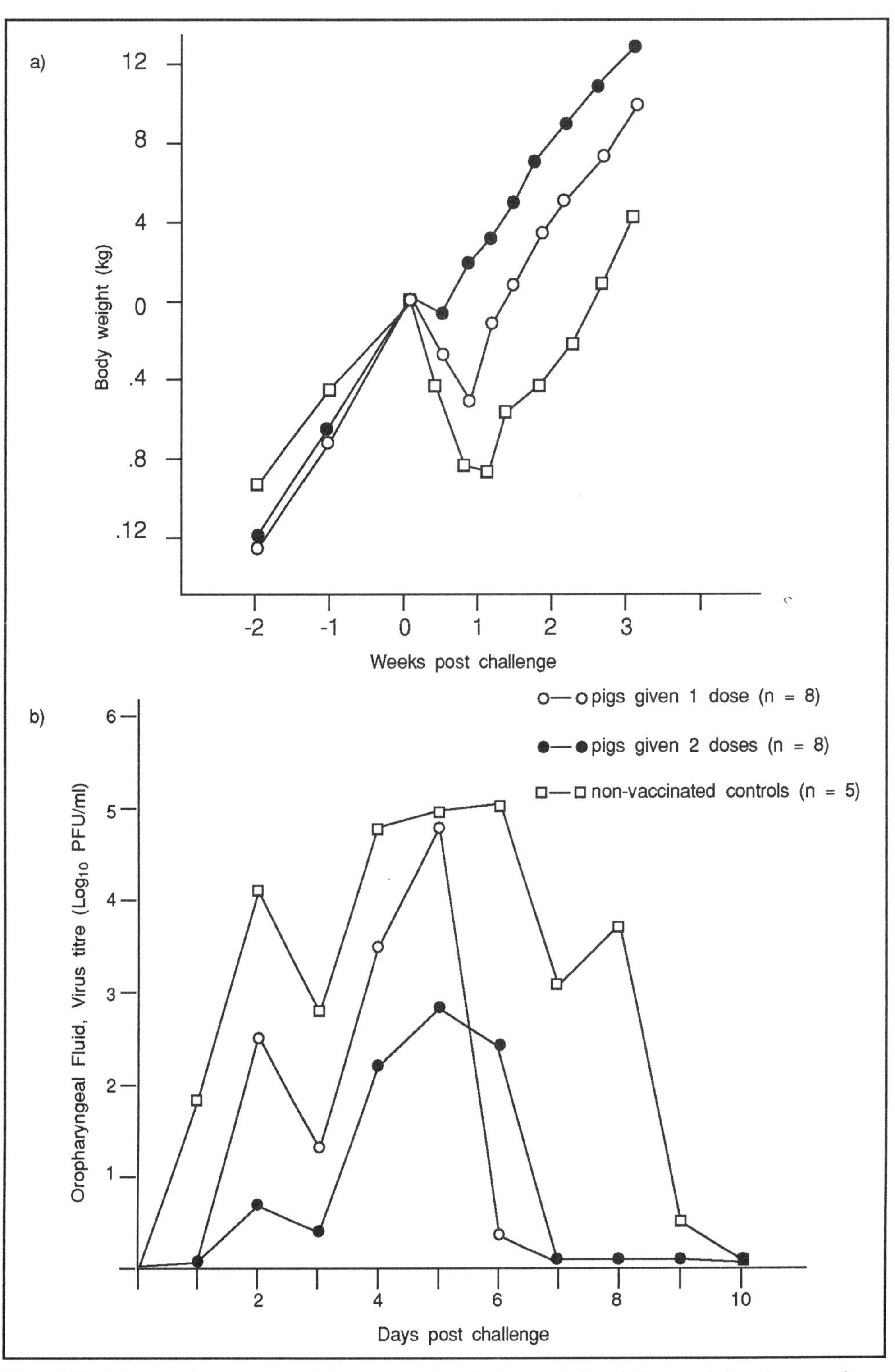

Figure 7.5 a) Mean weight curves and virus in secretions before and after challenge of pigs given vaccine A against Aujeszky's disease. (Data from van Oirschot and Gielkens (1987), Veterinary Quarterly 9 suppl. 1 37-49).

Π Table 7.3 and Figures 7.5 a) and b) show results of tests of a vaccine against Aujeszky's disease. Summarise the main results of these tests.

Vaccination of pigs against Aujeszky's disease has several effects. After infection with a virulent virus strain, vaccination:

- prevents mortality;
- reduces the severity of clinical symptoms;
- reduces virulent virus shedding (both quantity and duration);
- increases the viral dose required to infect pigs;
- reduces the incidence of induction of reactivation/latency of virulent virus.

Thus pigs vaccinated against Aujeszky's disease can still develop clinical symptoms and shed virulent virus after infection, which may result in virus transmission to pigs in contact.

On the other hand, pigs given a single vaccination against swine fever (parvovirus) do not become ill nor shed virulent virus after challenge infection. A great difference thus exists in efficacy of these two vaccines. However, this cannot be fully explained by different vaccine properties; it is also caused by the pathogenesis and immunology of the respective viral infections. Swine fever is a systemic infection, characterised by viraemia and virus distribution throughout the host's tissues, whereas an Aujeszky's disease is more confined to the respiratory tract and brain. In addition, the immunity against swine fever is lifelong, whereas that to Aujeszky's disease is relatively shortlived. The mechanisms underlying this immunological discrepancy are far from clear. Although much progress has been made in immunology during the last decade, the precise processes leading to immune responses against infections and vaccinations are largely unknown. This lack of basic knowledge in the field of immunology hampers rapid progress in the development of modern vaccines.

Vaccines against Aujeszky's disease should also prevent mortality in the newborn piglets due to the disease. The sows, which are vaccinated two to three times a year, produce antibodies against the virus and transmit these antibodies via their colostrum to their offspring. These maternally derived antibodies protect the highly susceptible newborn piglets against developing severe neurological signs after viral infection. Therefore, vaccines are examined for their ability to prevent neurological signs or death in four week-old piglets that are born to at least twice vaccinated sows. Together with pigs of the same age, but born to unvaccinated sows, they are challenged with virulent Aujeszky's disease virus and the incidence of neurological signs and death are recorded in both groups. A vaccine that, indirectly, prevents mortality in 80% of piglets, whereas 80% of the piglets born to unvaccinated sows die after the challenge infection, can be considered efficacious.

Maternal vaccination to immunise offspring passively, as described above, is also utilised for prevention of neonatal diarrhoea caused by *Escherichia coli* infections in cattle and pigs.

7.5.2 Safety

safety

The safety of a vaccine is primarily examined in the target animal. There is no common experimental procedure for evaluating safety of vaccines. The animal experiments that must be carried out are dependent on the features of the vaccine, for instance whether it is a live or killed vaccine, and on the characteristics of the disease to be prevented by vaccination. However, there are some general safety requirements for veterinary vaccines, which will now be described.

In reading through this section, it might be helpful to make a list of the safety tests that should be carried out on a vaccine.

contaminating agents

The vaccine must not contain contaminating agents, such as bacteria, fungi, or viruses. Absence of bacteria, fungi, and mycoplasmata has to be checked by *in vitro* procedures. To control for extraneous viruses, the target animal is inoculated at least twice with multiple doses of the vaccine, and the sera, collected from these animals two weeks after the last inoculation, must be negative for antibodies against various viruses.

sign of disease

The vaccine must not cause severe signs of disease in the target animal. Usually multiple vaccine doses are inoculated into young animals which are then monitored for the development of clinical signs, rise in body temperature and reactions at the site of inoculation. An uninoculated group of animals of the same age, housed under the same conditions, serves as control. Mild and transient clinical signs are generally acceptable. Sometimes, particularly in the case of live vaccines, the experimental animals first undergo corticosteroid treatment to stimulate stress conditions.

genetic stability of live vaccines

More specific safety requirements relate to live and killed vaccines. The genetic stability of a live vaccine is often assessed by conducting a number of serial passages in the target animal. The original virus and the virus collected during the last animal passage are compared for virulence characteristics in animals. The last passage virus must not be more virulent than the original vaccine virus.

transmission of vaccine viruses

A high degree of transmission of vaccine virus to contact animals is generally not acceptable. However, in the poultry industry this feature is not considered to be as negative as in the swine industry.

affect on fertility

The vaccine should not influence the fertility of animals. For instance, sows that are vaccinated during pregnancy should not farrow smaller litters than unvaccinated sows. In contrast to the other safety experiments, these 'fertility experiments' are usually done in the field, because many animals are involved.

inactivation of killed vaccines

For killed vaccines it is crucial that the inactivation procedure is effective in completely destroying the infectivity of the agent. This must be thoroughly controlled. For instance, during the production of foot-and-mouth disease vaccines several *in vitro* and *in vivo* experiments are performed to verify the absence of infectious virus in the vaccine. In one of these experiments, three cattle are inoculated in the tongue, and they obviously must not develop signs of foot-and-mouth disease.

inflammatory reactions

The adjuvants incorporated in killed vaccines often give rise to local inflammatory reactions at the site of inoculation. Therefore, animals are vaccinated repeatedly and after each vaccination, generalised or local adverse reactions are monitored. Severe side-effects are unacceptable.

7.6 Biotechnologically designed vaccines

The introduction of recombinant DNA and hybridoma technology has revolutionised research on the development of vaccines. Whereas conventional vaccines were to a certain degree developed by 'trial and error' procedures, biotechnology offers new strategies for the well-defined engineering of vaccines. However, before vaccines can be designed, more knowledge is required on the nature of microbial immunogens, the mechanisms underlying virulence of microbes, the (molecular) pathogenesis of an infection and the immunological processes responsible for induction of protective immunity. Biotechnology has created tools for the detailed studies of these topics.

identification of immunogenic components

Various methods are available to identify the immunogenic proteins of micro-organisms and to map the genes encoding these proteins. For example, monoclonal antibodies (MAbs) are produced against a virus and their protein reactivity is determined in a radioimmuneprecipitation assay. These MAbs are injected into animals that are subsequently challenged with a virulent virus. The MAbs that best protect the animal against disease are probably directed against the most immunogenic protein.

identification of genes coding for immunogenic proteins

Genes encoding the immunogenic proteins can be identified in several ways. For example, microbial DNA is sheared to small fragments by sonication, the DNA fragments are subsequently isolated and cloned into an expression vector. The expression products are screened in a Western blot with MAbs of which the protein-reactivity is known. The DNA cloned in the vector is then mapped to known restriction fragments of the genome by Southern blotting. Thus, the genes coding for a certain (glyco) protein may be localised on the genome.

antigenic determinants or epitopes

The ability to clone the genes encoding the immunogenic proteins and determine their nucleotide sequence has allowed the determination of the amino acid sequence (primary structure) of these proteins. Once the primary amino acid sequence is known several calculation methods (algorithms) can be used to predict the location of the immunogenic regions ('antigenic determinants' or 'epitopes') on a protein. However, such methods often do not indicate the region of interest.

A method which does not predict, but directly identifies the epitopes on a protein is the PEPSCAN method (Figure 7.6).

The PEPSCAN method involves the sequential synthesis of overlapping sets of peptides (in Figure 7.6 we have shown nona-peptides) corresponding to the sequence of the protein. The reactivity of each peptide with particular antibodies is then determined using an enzyme linked immunosorbent assay (ELISA). The example depicted in Figure 7.6 is from a protein of a transmissible gastro-enteritis virus.

With the PEPSCAN method, the biological relevance of epitopes can be determined by synthesising certain peptides, injecting these into animals and analyzing whether the animal's antiserum recognises the protein and neutralises the virus.

Studies such as those described above have greatly enhanced our understanding of the molecular nature of immunogenicity and virulence of micro-organisms. This knowledge provides the basis for new approaches to develop vaccines.

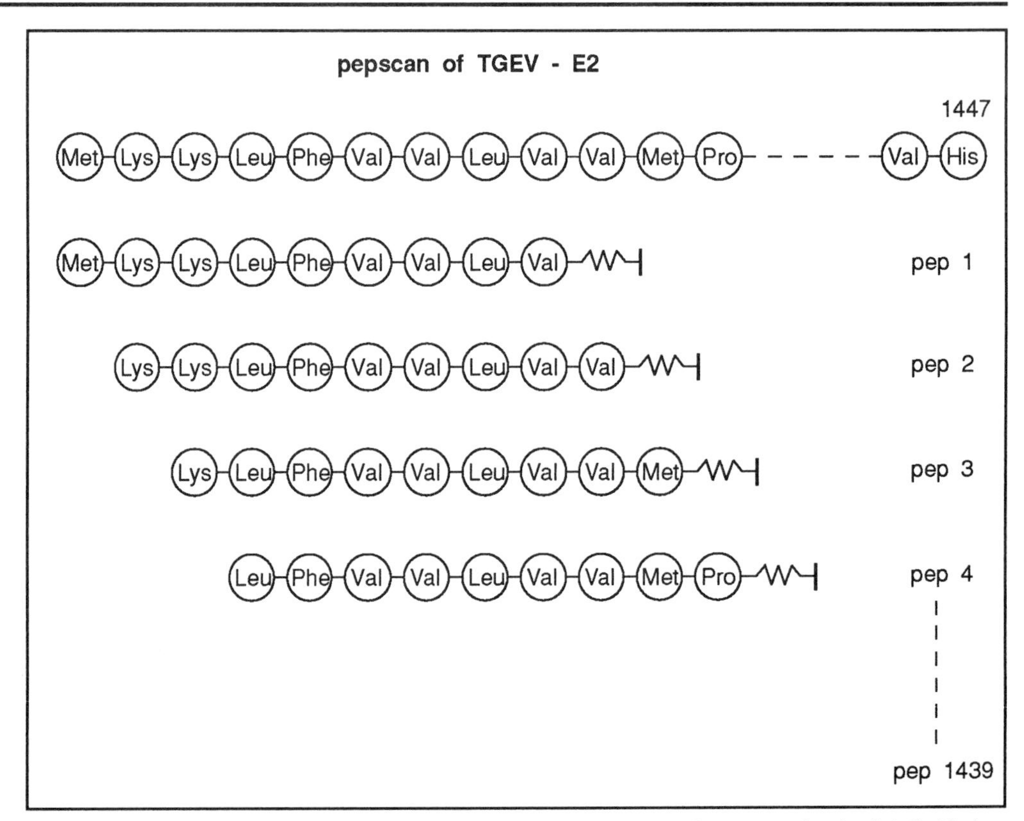

Figure 7.6 The PEPSCAN method for identification of epitopes on a protein (see text for details). By kind permission of R.H. Meloen.

Once it is known which of the microbial proteins (or portions of the protein) are involved in inducing immunity and, if the nucleotide sequence of the encoding gene is available, a number of ways are open to design 'biotech' vaccines. These are summarised in Table 7.4 and the various approaches are discussed in the next sections. It is also convenient to categorise such vaccines as 'live' and 'killed'.

'Killed' vaccines
1. Subunit vaccines
2. Recombinant DNA vaccines -polypeptide expression in bacteria, yeast, viruses or mammalian cells.
3. Synthetic peptide vaccines
4. Anti-idiotype vaccines
'Live' vaccines
1. Deletion mutant vaccines
2. Reassortant vaccines
3. Vaccinia vectored vaccines (vector vaccines).

Table 7.4 Modern approaches for vaccine development.

Π List as many alternatives for the terms 'live' and 'killed' vaccines as you can, then check it against our alternatives described below.

Alternatives for 'live' are replicating, attenuated, modified-live, and infectious. Instead of 'killed' vaccines the following terms can also be used: non-replicating, inactivated, non-infectious. All of these terms are used in the literature. We will examine the strategies listed in Table 7.4 in Sections 7.7 and 7.8.

7.7 Killed vaccines

7.7.1 'Conventional' subunit vaccines

conventional subunit vaccines

Subunit vaccines are vaccines made up of components of the disease causing agent. An immunogenic protein of a virus, or a combination of immunogenic proteins can provide the basis for a 'conventional' subunit vaccine. Conventional in this case means that no recombinant DNA technology is involved. In fact, various bacterial vaccines now in use are so produced, for instance *E. coli* pili vaccines and tetanus toxoid vaccines. The immunogenic proteins must be extracted from cells or harvested from cell culture fluids, concentrated and subsequently purified. A vaccine against Aujeszky's disease that is comprised of two immunogenic glycoproteins of the virus is also on the market. A promising approach to subunit vaccines are iscom vaccines. The word iscom is derived from ImmunoStimulating COMplex. Immunogenic proteins are adsorbed onto Quil A, giving rise to complexes carrying the protein in a multimeric form on their surface. In this form, Quil A serves as an adjuvant. The production of iscoms on a scale large enough for vaccine production is still a technical problem that remains to be resolved.

iscom vaccines

7.7.2 Recombinant DNA vaccines

recombinant DNA vaccine

Once the gene coding an immunologically important protein is identified, it can be inserted into and expressed in a variety of biological systems, which can be grown in culture, yielding high quantities of the immunogenic protein or portions thereof. The polypeptides produced can form the basis for a subunit vaccine. Such polypeptides are frequently used as research tools and are occasionally applied as a diagnostic tool. As biological expression systems (or host cells) the following are available: bacteria; yeast; viruses and mammalian cells.

Bacteria or prokaryotes

The most commonly used bacterium is *Escherichia coli*. By employing standard recombinant DNA methods the gene of interest is inserted into a plasmid vector, which is then used to transform the host cell. Bacteria producing the desired polypeptide are identified, selected and cultured. The polypeptide is extracted from host cells, concentrated, purified, and used as the basis for a vaccine.

pili

In 1982, a recombinant DNA *Escherichia coli* vaccine to prevent neonatal diarrhoea in pigs was launched. The conventional vaccines contained purified preparations of a surface antigen of the bacterium called pili. Pili are involved in the adsorption of the bacteria onto mucosal epithelial cells. These pili used to be produced by wild-type strains of *E. coli*, which yielded relatively low amounts of the desired product. However, by cloning the pili-genes downstream of strong promotors (reshuffling of genes) the yield could be increased at least ten-fold. Such engineered strains are currently used to produce pili, that are subsequently purified and incorporated into a vaccine.

feline leukaemia

In 1988, a recombinant DNA vaccine against feline leukaemia was marketed. This vaccine contains an immunogenic viral protein that has been produced in *Escherichia coli.* It was the first commercially available recombinant DNA veterinary vaccine to be expressed in a prokaryote. It took seven years to develop the vaccine.

Feline leukaemia is a disease of cats caused by a retrovirus. The genome of feline leukaemia virus consists of a 60-70S single stranded RNA comprising a gag gene, encoding for the core protein, a pol gene coding for polymerase, and an env gene encoding the gp70 and p15 envelope proteins. Previous studies indicated that the gp70 protein is the most immunogenic viral protein and responsible for adsorption onto the cells of cats by binding to the cell receptor.

Π Why might these attributes of the gp70 protein be valuable in a vaccine?

Antibodies against the critical epitope on gp70 will prevent the virus adsorbing onto the cat cells. The gp70 is therefore the protein of choice to be incorporated into a vaccine. The viral gene for gp70 was cloned into a plasmid vector and introduced into *Escherichia coli.* After bacterial growth at 37°C, the temperature is increased to 42°C to activate regulation signals for the production of the gp70. In bacteria the expressed polypeptide is not glycosylated, hence the molecular weight is approximately 45 kilodalton rather than 70 kd. This 45 kD protein carries the structure responsible for binding to the cell receptor. The polypeptide is extensively purified and adsorbed onto alhydrogel and Quil A. This vaccine has been found safe and afforded protection, although not complete, in cats against challenge with a field strain of feline leukaemia virus.

Yeast

Baker's yeast, *Saccharomyces cerevisiae,* is mostly used to synthesise heterologous proteins. The advantage of this micro-organism is that it can be efficiently produced on a large-scale. In contrast to bacteria, yeast and other fungi are able to glycosylate proteins and perform proteolytic processing. This is of particular importance if the glycosylation is needed to produce the correct antigenicity and thus stimulate the production of antibodies which recognise the naturally occurring structure of the infective agent.

hepatitis B vaccine

The recombinant DNA hepatitis B vaccine, which hitherto represented the biggest success of biotechnological vaccine development, and was marketed in 1986, is produced in yeast. The gene coding for HbsAg (a surface antigen of hepatitis B) is inserted into yeast cells and is expressed as particles that are morphologically quite similar to the particles found in the blood of carriers. In contrast, the gene products expressed by *Escherichia coli* did not assemble into the form of the plasma particles, and are less immunogenic.

The 'conventional' vaccine for hepatitis B originates from plasma of human carriers of hepatitis B virus. This plasma contains surface antigens (HbsAg) assembled into 22 nanometer noninfectious particles. These particles are extensively purified and form the basis of the vaccine. It should be obvious that such a vaccine carries several disadvantages.

A case study describing the details of both conventional hepatitis B vaccine and the yeast produced recombinant DNA hepatitis B vaccine is described in the BIOTOL text 'Biotechnological Innovations in Health Care'.

SAQ 7.5

Make a list of some disadvantages relevant to vaccines derived from human blood.

Viruses

Large DNA viruses are the prime candidates for being used as expression systems for foreign genes. Vaccinia virus, which can replicate in many domestic animals and man, will be dealt with in Section 7.8.3 on live vaccines.

Mammalian cells

Many viral proteins have already been expressed in various *in vitro* cultured cells. The yield of the foreign protein is relatively low in comparison with that in bacteria or in yeast. Cells expressing recombinant proteins are often used in research on functions of these proteins. A disadvantage of mammalian cells for vaccine production is that cell lines may be transformed by oncogenes, thus having oncogenic potential. The attraction of using mammalian cells is that they carry out glycosylation and usually produce the desired antigen in the correct conformation. Mammalian cells are however difficult to cultivate on a large scale and this currently limits their general application for producing killed vaccines on a commercial scale.

No particular biological expression system deserves preference for producing vaccines. Each attempt to express immunogenic polypeptides must be evaluated on an individual basis.

7.7.3 Synthetic peptide vaccines

synthetic peptide vaccines

This development aims at utilising solely the immunogenic fractions of a protein, usually oligopeptides, as a basis for a synthetic peptide vaccine. To synthesise a peptide the amino acid sequence which comprises an antigenic determinant or epitope needs to be determined. If the nucleic acid sequence of the encoding gene is known, the amino acid sequence can be deduced. A number of predictive approaches can then be employed to identify potential epitopes recognised by antibodies (B cell epitopes), for instance the hydrophilicity plot. This is based on the assumption that a hydrophilic region of amino acids is more likely to be on the surface of a micro-organism and thus recognised by antibodies than a hydrophobic region. In contrast with the predictive approaches, which often fail to indicate an epitope, the use of the PEPSCAN technique enables us to define exactly the amino acid sequence constituting an epitope (Figure 7.6).

Briefly, the PEPSCAN method involves the sequential synthesis of overlapping sets of peptides corresponding to the sequence of the protein. The reactivity of such peptides with antibodies raised against the native protein or against the whole micro-organism is determined in an enzyme - linked immunosorbent assay. This method mainly detects continuous epitopes. Continuous or linear epitopes are determined by their linear amino acid sequence, whereas discontinuous or conformational epitopes are formed by amino acid residues far apart on the primary structure but brought together by the folding of the protein (Figure 7.7). Most protein epitopes recognised by B cells are conformational. However, it is difficult to identify discontinuous epitopes and consequently to include them in chemically synthesised peptide vaccines. Work on foot-and-mouth disease virus, described in the next few paragraphs, illustrates some problems encountered when developing synthetic peptide vaccines.

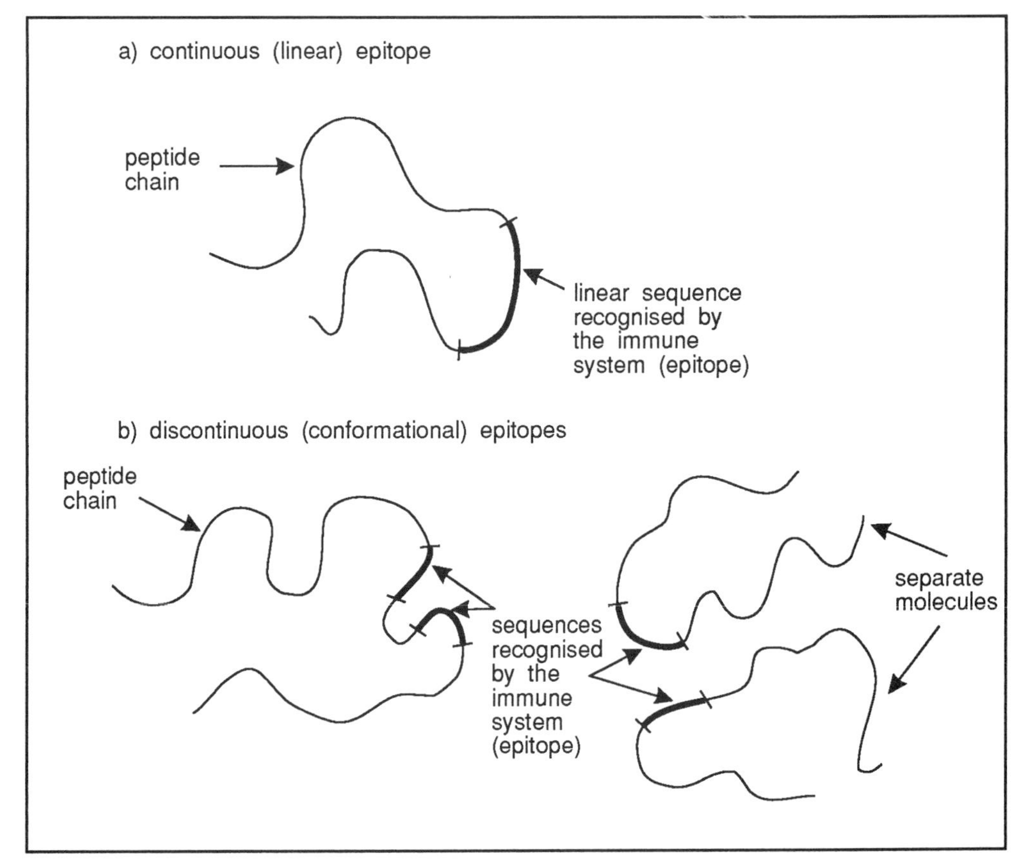

Figure 7.7 A schematic representation of a) continuous (linear) epitopes, b) discontinuous (conformational) epitopes.

The background research, and some of the problems encountered in the development of a subunit vaccine against foot-and-mouth disease based on expression of (portions of) virus protein 1 (VP1) in bacteria, or by chemical synthesis, are now described.

It was first established in 1969, that enzymatic cleavage *in situ* of VP1, which is one of the four structural proteins of the virus, was accompanied with loss of immunogenicity. This suggested an important role for VP1 in induction of immunity. Indeed, the VP1 isolated from the virus particle was shown to elicit neutralising antibodies and to induce protective immunity in swine and cattle. However, compared with the activity of the intact virus particle the isolated VP1 protein is only weakly immunogenic.

Π Can you suggest what these observations mean?

These observations indicate that immunogenicity is dependent on the integrity of the virus particle.

VP1 or fragments of VP1 have been made by recombinant DNA technology, using an *Escherichia coli* expression system. In the first study, the host cells expressed one to two million copies per cell of a VP1 fusion protein. The extracted and purified fusion protein elicited neutralising antibodies and protective immunity against type A12 foot-and-mouth disease virus in both swine and cattle. In 1981, US Secretary of Agriculture John R. Block described this achievement as 'the first production through gene splicing of an effective vaccine against any disease in animals or humans'. The Wall Street Journal of 19 June, 1981 announced that 'This breakthrough can mean annual savings of billions of dollars and an increase in the supply of meat'. However, these optimistic views have not yet been fulfilled. Several problems arose regarding the development of a recombinant DNA vaccine for foot-and-mouth disease produced in prokaryotes. For instance, such vaccines developed for another type of foot-and-mouth disease virus (type 0_1, which is one of seven virus serotypes) did not induce satisfactory immunity, probably due to the low intrinsic immunogenicity of type 0_1 virions.

At about the same time it was reported that a single dose of a synthetic peptide vaccine, containing amino acid residues 141-160 of VP1 linked to keyhole limpet haemocyanin as a carrier protein elicited neutralising antibodies and protective immunity in guinea pigs. However protection in cattle, the target animal, was much less satisfactory. One problem encountered was that coupling of the peptide to a carrier protein did not result in a uniform product with a known location and configuration of the peptide on the carrier protein. However, a polymerised peptide, without a carrier protein, also proved to be immunogenic. The somewhat disappointing results with synthetic peptide vaccines may be explained by the differences in configuration of the synthetic peptide and of the same peptide when it forms part of the virus particle.

Therefore, attempts were made to produce highly immunogenic peptides by expressing a peptide (amino acid residues 137-162) as a fusion protein attached to the N-terminus of β-galactosidase. Fusion proteins with two or four copies of the peptide were superior to the one copy construct or the peptide alone in evoking neutralising antibodies. It is likely that the dimeric or tetrameric construct has an altered configuration, which may be the key factor in the enhanced antibody response. The immunogenicity of the peptide could be greatly enhanced by expressing it as a part of the hepatitis B core protein in vaccinia virus. The core particles are morphologically similar to unmodified hepatitis B core particles. Very low amounts of these particles elicited protection against challenge infection in guinea pigs. The immunogenicity of foot-and-mouth disease virus peptide/hepatitis B core antigen fusion protein approached that of conventional inactivated foot-and-mouth disease vaccines.

Π Comparative immune responses of three species to foot-and-mouth disease synthetic viral protein are shown in Figure 7.8. What are the different levels of response and their implications?

In contrast with their performance in guinea-pigs, foot-and-mouth disease virus peptides are poor immunogens in the target species, cattle and pigs (Figure 7.8). From this finding and others, it has become clear that a peptide is only immunogenic in recipients whose histocompatibility proteins recognise it. It appears that uncoupled peptides must contain epitopes which react with helper T cells and Ia (histocompatibility antigens) antigens in addition to B cell epitopes to overcome genetically restricted non-responsiveness. For instance, H-2_d mice, which are non-responders to the immunogenic peptide of foot-and-mouth disease virus do respond to the same sequence if it is linked to defined helper T-cell determinants from

Ia antigen

ovalbumin or sperm whale myoglobin, thus overcoming genetic restriction of the immune response.

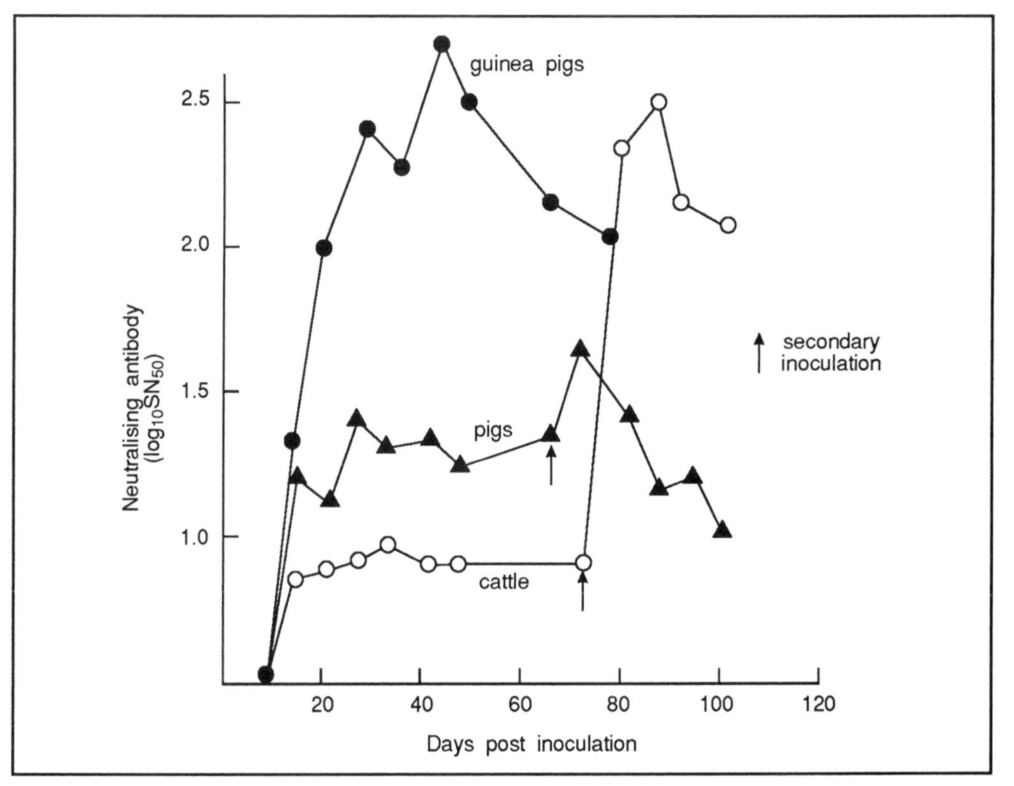

Figure 7.8 The neutralising antibody response in natural hosts and laboratory animals to the free 137-160 cys peptide from VP1 of foot-and-mouth disease virus, serotype 0_1. (Data from Francis *et al* (1988), Proc. Cold Spring Harbor Lab. p1-7).

Although a decade has elapsed since the first published report of a putative recombinant DNA vaccine for foot-and-mouth disease, no modern subunit vaccine is expected to enter the market in the next five years. Although this may be interpreted as rather disappointing, recent developments on enhancing peptide immunogenicity and overcoming the genetic restriction of the immune response still offer prospects for future biotech foot-and-mouth disease virus vaccines, be they recombinant DNA products or synthetic peptides.

7.7.4 Anti-idiotype vaccines

anti-idiotype vaccines

This strategy to develop anti-idiotype vaccines is based on the network theory that Jerne postulated in 1973. The essential element of this theory with regard to vaccine development is that the antigen combining site (idiotype) of an antibody can function as antigen and therefore elicits antibodies directed against the idiotype, thus termed anti-idiotype antibodies. Some anti-idiotype antibodies only recognise antibodies, whereas others (the internal image anti-idiotypes) have the same tertiary configuration as the antigen to which the first antibodies are evoked, and will therefore mimic the behaviour of the antigen. Therefore they may give rise to antibodies that are directed against the original antigen, and thus act as a vaccine.

Π Can you think of any advantages of anti-idiotype vaccines?

Monoclonal anti-idiotypes are safe to prepare and available in large quantities. They provide epitope-specific reagents and provide structures which mimic the behaviour of the tertiary configuration of the native antigen in a way which synthetic peptides might not always be able to do. The anti-idiotypes can also mimic nonprotein structures such as carbohydrates which cannot be produced directly by gene cloning.

7.8 Live vaccines

7.8.1 Deletion mutant vaccines

deletion mutant vaccines

This approach is feasible for viruses or bacteria which carry genes that are non essential for their replication. It will be exemplified by deletion mutant vaccines that have been engineered for Aujeszky's disease virus. Aujeszky's disease is caused by a herpes virus to which the pig is the natural host.

The genome of Aujeszky's disease virus consists of a linear double-stranded DNA molecule approximately 150 kilobases in size. It is composed of an unique long sequence and an unique short sequence. The mature virion carries about 50 proteins. The genes coding for the glycoproteins gII, gIII, gH, gX, gp50, gp63, and gI, for an 11K protein and for thymidine kinase have been mapped on the genome (Figure 7.9). Viruses that fail to express gIII, gX, gp63, gI or the 11K protein still replicate *in vitro*, indicating that these genes are not essential.

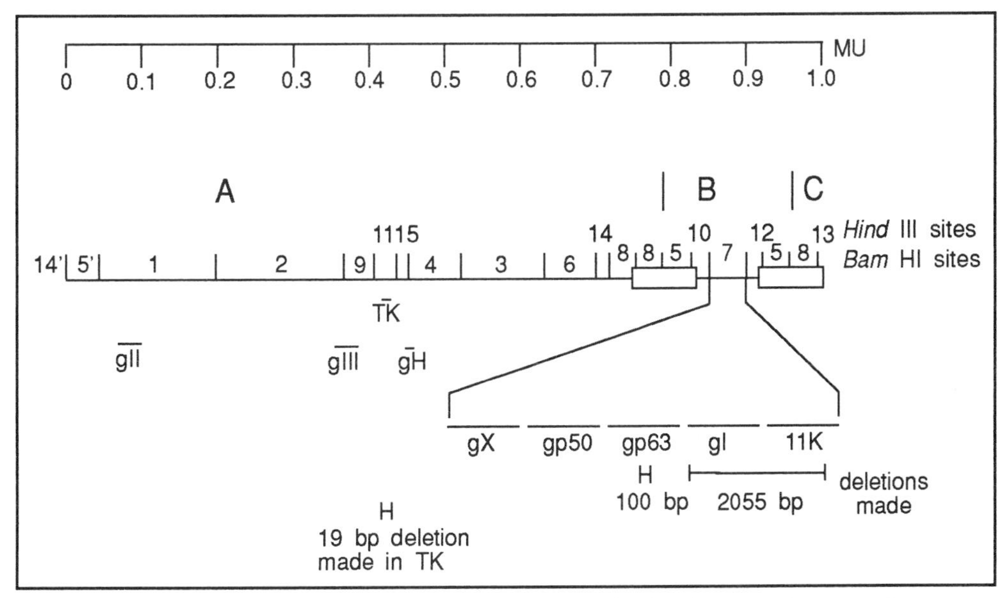

Figure 7.9 Structure of the genome of Aujeszky's disease virus, showing the mapped genes coding for a series of glycoproteins, gII, gIII, thymidine kinase (TK) and other genes. The construction of the deletion on mutants to produce vaccines is also indicated. The removal of 19 bp from the thymidine kinase gene leads to the production of inactive thymidine kinase.

virulence

Virulence is a feature of a micro-organism that indicates to which degree a certain strain produces disease, when infecting a susceptible host. The virulence of Aujeszky's disease virus is multigenically controlled, (ie products of multiple genes are involved in the expression of virulence). The non-essential gIII, gp63, gI, and the thymidine kinase enzyme play a role in virulence, whereas gX does not. Several research groups have constructed deletion mutants by deleting non - essential virulence genes. Recombinant DNA technology has been applied to engineer the deletion mutants. Generally, the genes involved were first inserted into plasmid vectors, that subsequently were used to engineer deletions. Cultured cells were transfected with the plasmid carrying the deleted gene and infected with the wild-type virus resulting in recombinants, which were subsequently selected for the phenotype of choice.

There are three kinds of deletion mutant vaccines for Aujeszky's disease commercially available. None of these vaccines expresses thymidine kinase. In addition, the vaccines are deficient in the expression of either gIII, or gX, or gI. Thus, these vaccines have at least two deletions; in two vaccines two virulence genes are inactivated, whereas the third vaccine fails to express one gene involved in virulence.

Π List some advantages of deletion mutant vaccines against Aujeszky's disease compared with conventional vaccines?

These deletion mutant vaccines are safe and at least as immunogenic as conventionally attenuated Aujeszky's disease vaccines. A great advantage of deletion mutant vaccines is that the antibody response they elicit can potentially be differentiated from the antibody response after infection with a wild-type virus. This feature makes it possible to combine vaccination schemes with programmes to eradicate infectious animal diseases. For example a deletion mutant vaccine in which gene gX has been deleted will not cause animals to produce anti-gX antibodies. Wild type viruses will however contain gX genes and thus animals infected with these will produce anti-gX antibodies. We can therefore identify animals infected with the wild type virus. In the section on diagnostics more attention will be paid to this aspect.

Genes may also be inactivated by inserting pieces of foreign DNA, for instance restriction enzyme linkers, into a gene. The immunogenicity of deletion mutants may be further improved by enhancing the expression of the most immunogenic proteins of the disease causing agent.

7.8.2 Reassortant vaccines

reassortant vaccines

This approach for modern vaccines does not actually involve *in vitro* DNA recombination. Reassortant vaccines can be prepared from viruses that have segmented genomes, ie influenza and rota viruses. Their genetic information residies on eight to ten RNA segments. By simultaneously infecting cell cultures with two different virus types, through a process of recombination, viruses may emerge that carry for instance one segment of RNA of one virus type and seven RNA segments of the other virus type. This 'reassorting' of genes is applied in the production of influenza virus vaccines.

This is an empirical approach and depends upon having suitable screening procedures to identify the desired 're-assorted' strain.

7.8.3 Vaccinia vectored vaccines

vaccinia virus

One of the most studied and most promising approaches to the design of new vaccines is the use of vaccinia virus as a vector, or 'carrier', or as an expression system for genes of heterologous micro-organisms. Thus, the virus that Jenner used to perform his immunisation experiments almost two hundred years ago, is now a promising tool for biotechnologists attempting to engineer new vaccines. However, not only vaccinia virus may be used as a vector, but also other large DNA viruses, such as other poxviruses, herpes viruses and perhaps adeno viruses. Use Figure 7.10 to help you follow the description below.

The gene coding for an immunogenic protein of a microbe is inserted into a plasmid vector containing a vaccinia virus promotor flanked by vaccinia DNA from a non-essential gene, usually from the thymidine kinase gene. This plasmid is transfected into cells already infected with a vaccinia virus strain. Homologous recombination between the non-essential vaccinia DNA of the plasmid and of the virus results in insertion of the heterologous gene into the vaccinia virus genome, rendering the recombinant thymidine kinase negative. This absence of thymidine kinase expression is used to select recombinants. Such a selection system is necessary because the frequency of recombination is usually less than 0.1%. Vaccinia virus carrying more than 25 000 nucleotides of foreign DNA is still stable and replicates normally *in vitro*. Consequently, vaccinia virus can be used to express multiple foreign genes. To stimulate the immune response elicited by the heterologous genes, the genetic information for interleukin 2 may also be inserted. Generally, the foreign gene is authentically expressed, and the immunogen is presented in a manner identical to that during an infection.

In this procedure, a chimaeric plasmid is constructed in such a way that the coding sequences are inserted downstream of a vaccinia virus promoter sequence. This results in an inactive TK gene. Cells infected with vaccinia virus are transfected with the chimaeric plasmid. Recombinant viruses are selected by using bromodeoxyridine (Bdllr). TK^- recombinants are insensitive (resistant) to Bdllr whereas TK^+ recombinants are inactivated by this reagent.

vaccinia vectored vaccines

Numerous animal experiments have been conducted with experimental vaccinia vectored vaccines to evaluate safety and efficacy. The results are most promising, in that a good level of protective immunity was usually induced.

Advantages of a vaccinia vectored vaccine are:

- its potential to induce a high level of protection, because it is a replicating vaccine;
- its potential to be used as a multivalent vaccine vector;
- it is physically stable, and therefore after freeze-drying no 'cold chain' is required to maintain its infectivity. This is a crucial feature with regard to vaccinations to be carried out in underdeveloped countries;
- it is genetically stable and as a result no reversion to virulence will occur;
- it is cheap, because it is easy to produce. With the calf lymph method, approximately one million doses can be produced per calf;
- it is easy to administer with a bifurcated needle or a skin jet.

There are also disadvantages of vaccinia vectored vaccines:

- when an immunocompromised host is vaccinated, it may develop generalised vaccinia, eczema, or encephalitis;

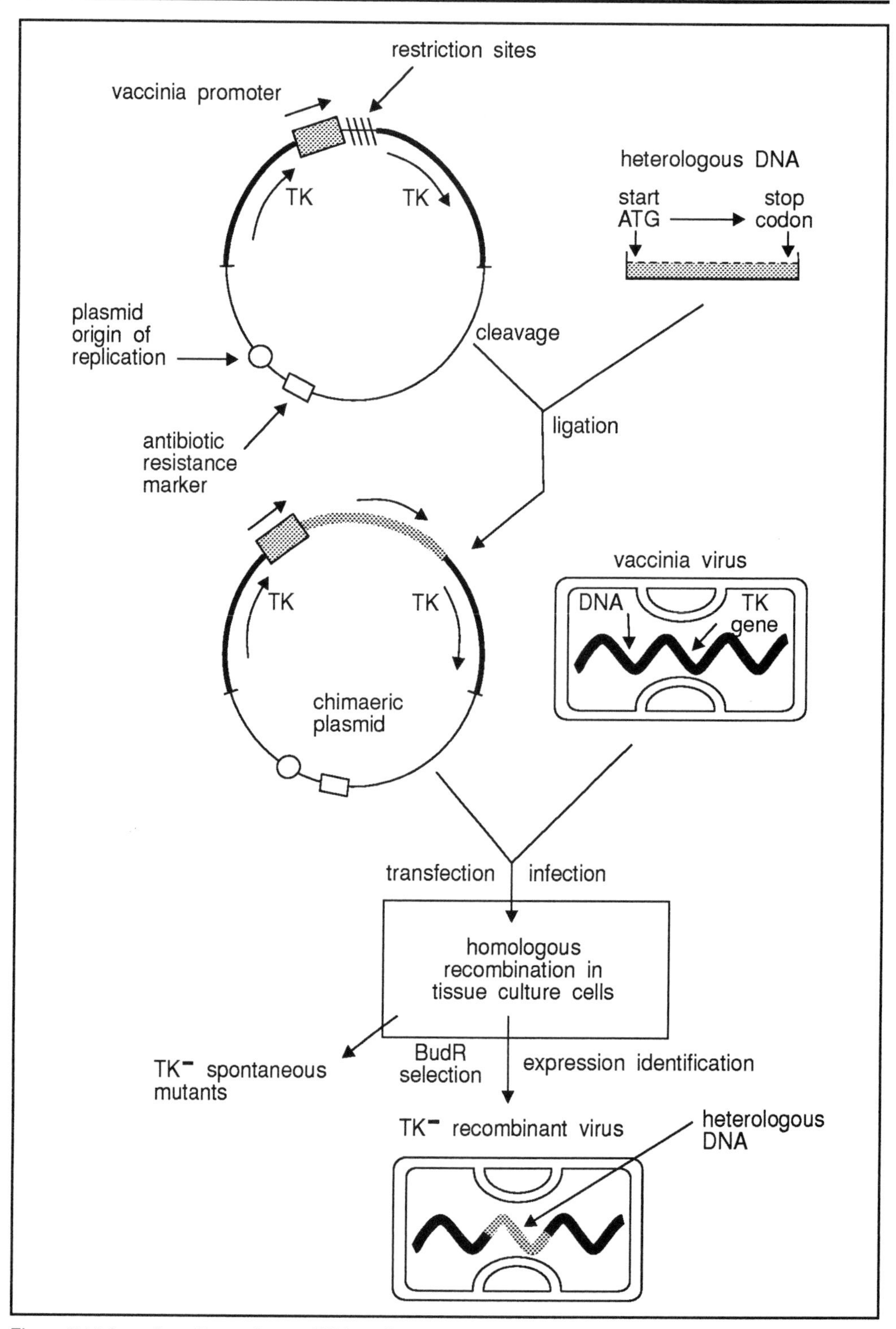

Figure 7.10 Insertion of heterologous DNA coding sequences into vaccinia virus by insertional inactivation of the virus thymidine (TK) gene. (After Esposite and Murphy 1989, Advances in Veterinary Science and Comparative Medicine 33 195-247 Academic Press).

- vaccinated animals must be carefully handled in the short term, because vaccinia virus is contagious to man. Also products from animals vaccinated with vaccinia vectored vaccines may be contaminated with vaccinia virus;
- the vaccinia vectored virus may infect new hosts if the expressed foreign protein is involved in virus adsorption and penetration.

None of the experimental vaccinia vectored vaccines is currently commercially available. At present, field experiments are in progress in Belgium to evaluate efficacy and safety of a vaccinia vectored rabies vaccine.

Other DNA viruses have also been used as vectors for heterologous genes. for example, a glycoprotein I - negative deletion mutant of Aujeszky's disease virus has been successfully used as vector for the immunogenic E1 protein of swine fever virus. Immunisation of pigs with this herpes virus vectored vaccine induced a high level of protection against both swine fever and Aujeszky's disease. In this case the vector thus functions not only as a carrier, but also as a live vaccine.

SAQ 7.6

List the five modern biotechnological approaches to vaccine development?

SAQ 7.7

Let us assume that virus X causes a major disease in cattle and that you are seeking to produce a vaccine for the disease. You are attempting to decide what type of vaccine to produce. Below is a series of factors which may contribute to your decision. Label those which favour the choice of an attenuated virus with an A, those which favour an inactivated (killed) virus with a B and those which favour using a purified antigen with a C.

1) Special requirement for an adjuvant.

2) A risk of 'reversion' of the virus into a virulent form.

3) A greater immunological response.

4) Lack of knowledge of the specific antigens produced by the virus.

5) Greater vaccine stability.

6) Greater ability to characterise the vaccine.

SAQ 7.8

A disease of sheep is known to be caused by viruses which can exist in a variety of different forms. All are enveloped viruses and their capsids (coats) consist of about 160 capsomers (coat proteins). They are also known to be double stranded DNA viruses. We know the following about these viruses.

1) The capsomer proteins from the different viruses show many similarities but antibodies so far raised against these proteins show little affinity for the proteins derived from other capsids.

2) All of the envelopes of these viruses contain a mixture of glycoproteins. Analysis of these glycoproteins using gel electrophoresis indicates many similarities between the envelopes of the various viruses (ie similar sizes). These glycoproteins have not been characterised further.

3) All the viruses so far characterised produce an identical protein (M1) which appears to be involved in the maturation of the viruses. Antibodies prepared against M1 isolated from one virus reacts with M1 isolated from another. M1 is not a component of the envelope or the capsid of the viruses. It is produced within the host cells and it does not appear to be essential for viral replication.

4) The gene for M1 has been mapped onto a short (2.5 kbp) fragment of DNA produced by treating viral DNA with the restriction enzyme *Eco* R1. Incubation of viral DNA with this enzyme only produces 3 fragments.

Using this information, choose from the list below, the strategy most likely to lead to the production of a vaccine for the disease.

Strategy 1 - Clone to the *Eco* R1 fragment carrying the M1 gene into an expression vector and use this to produce a lot of the maturation protein which can be used as a vaccine.

Strategy 2 - Continue the genetic mapping to trace the capsomer protein genes. Then isolate these and clone them in an expression vector in order to produce capsomer proteins to be used as vaccines.

Strategy 3 - Use *Eco* R1 specifically delete M1 gene. Use a re-constructed virus composed of the remaining two *Eco* R1 fragments as the vaccine.

7.8.4 Efficacy and safety of biotech vaccines

Basically, the aspects of efficacy and safety discussed already also hold true for 'biotech' vaccines. Thus, the main advantage of live vaccines is their efficacy and for dead vaccines it is their safety.

An additional aspect to be considered with regard to live recombinant DNA vaccines is the potential dangers of releasing such a product into the environment. Before such vaccines are registered, extensive safety assessment studies should be conducted. In particular, there seems to exist a great deal of resistance in using vaccinia vectored vaccine in a world now free from smallpox. Environmental release is under regulating control. We remind you of the EC-directive relating to this and discussed in Appendix 1. More information on efficacy and safety testing with vaccinia vectored vaccines are required before these vaccines can be approved for use in animals. Vaccinia virus has the potential for use as a vector mainly in developing vaccines for severe infectious diseases such as AIDS, malaria and leprosy for which vaccines are not yet available.

7.8.5 Advantages of biotech vaccines

Modern biotechnological approaches to the design of vaccine have certain potential advantages compared with conventional vaccine development:

- vaccines may be developed from micro-organisms that do not grow or show only limited growth in culture. In fact, this already has become true for hepatitis B virus;
- vaccines against parasitic diseases (eg malaria) may become a reality. It is virtually impossible to develop such vaccines by conventional means;
- large-scale culture of agents hazardous to humans, eg rabies virus, or of agents hazardous to animals, eg foot-and-mouth disease virus, can be avoided. Stringent containment procedures are therefore no longer required;
- incomplete inactivation of infectivity, which sometimes occurs with conventional vaccines, is no longer possible;
- contamination with extraneous agents is less likely to take place;
- biotech vaccines will probably contain lower amounts of tissue culture components and other immunologically active contaminants, which lessens the chance of adverse reactions to vaccination. For instance, chemical synthesis of peptide vaccines implies relative purity and safety;
- biotech vaccines generally are expected to be more stable and heat-resistant, so that low temperature storage would not be required;
- biotech vaccines will elicit an antibody response that is differentiable by serological means from an antibody response after infection. This is of great importance for control of infectious diseases. It may be expected that combinations of vaccines and companion diagnostic kits will increase in the market.

7.9 Modern diagnostic methods

A confirmed diagnosis of an infectious disease can usually not be made by the veterinary practitioner in the field, on the basis of the anamnesis, the course of the clinical symptoms, and the pathological lesions. However, these observations often lead to a tentative diagnosis that can be confirmed in a laboratory. A laboratory diagnosis is based on detection of the micro-organism, and/or detection of a specific antibody response against the micro-organism.

7.9.1 Detection of the micro-organism

The micro-organism may be isolated from samples collected from the living animal, eg saliva, or from tissues of a dead animal. Alternatively an antigen produced by the micro-organism may be detected by immunological techniques, such as immunoflourescence or immunoperoxidase techniques.

Π Make a list of some key differences between the processes of isolation of a virus and detection of a viral antigen, and their implications for diagnosis?

For isolation of a virus, cell culture is required, for antigen detection it is not. Virus isolation takes at least a few days, whereas antigen detection tests can be completed in one day. Therefore, in routine laboratories antigen detection tests are often preferred to allow quicker diagnosis.

DNA/RNA probes

Since biotechnological methods became available, it is possible to detect parts of the genome of a micro-organism in samples and tissues of infected hosts with DNA/RNA probes. These methods are based on hybridisation (specific binding) between labelled DNA/RNA molecules (probes) and DNA/RNA in the test samples. Hybridisation only occurs if the nucleotide sequence of the probe and the DNA/RNA in the test sample are complementary. Recently, the polymerase chain reaction (PCR) has been introduced as a highly sensitive and specific diagnostic test to detect DNA or RNA.

PCR

The PCR is a method which is able to amplify femtograms (10^{-6} micrograms) of DNA to micrograms of DNA in a few hours. The amplified DNA can then be detected by common hybridisation procedures. The PCR can only be used to detect genes from which the sequence is, at least partly, established.

The PCR is performed as follows (Figure 7.11). The first step is to synthesise two 'primers', which are oligonucleotides (approximately 20 nucleotides), that are complementary to the positive and negative strand of DNA that flank the DNA segment to be amplified. The DNA in the test sample is denatured at 94°C, then the primers are annealed (hybridised) to opposite strands of the DNA at 37-55°C, and subsequently the DNA is extended starting from the primers using a heat resistant DNA polymerase. The DNA-polymerase is purified from the thermophilic bacterium *Thermus aquaticus*. It can survive prolonged incubation at 95°C, and thus is relatively unaffected by the denaturation step. The primers direct the synthesis of DNA by the polymerase to proceed across the region between the primers. Since the extension products are also complementary to and capable of binding primers, a repeat cycle of denaturation, primer annealing and extension doubles the amount of DNA synthesised in the previous cycle. Repeated cycles (eg 25-35 times) results in the exponential accumulation of the specific DNA fragment.

monoclonal antibodies in diagnosis

Another advance in diagnosis is the application of monoclonal antibodies (MAbs). For example, swine fever is diagnosed by detection of antigen in sections of tonsils of killed pigs. Antigen is made visible under the fluorescence microscope by reaction with anti-swine fever virus IgG that is conjugated to fluoresceine isothiocyanate. The IgG is purified from an antiserum that is raised by first vaccinating and then challenging pigs with a virulent virus. However, the immunofluorescence test also detects antigen from the closely related bovine virus diarrhoea virus. Because, in the framework of a national eradication programme, all pigs from a herd infected with swine fever virus must be killed and pig herds infected with bovine virus diarrhoea virus are not slaughtered, it is crucial to be able to differentiate between these virus infections. MAbs have been produced and selected that recognise swine fever virus antigen and do not react with bovine virus diarrhoea virus antigen. Thus, the use of these MAbs on tonsil sections of suspected pigs enables us to clearly differentiate between the two infections.

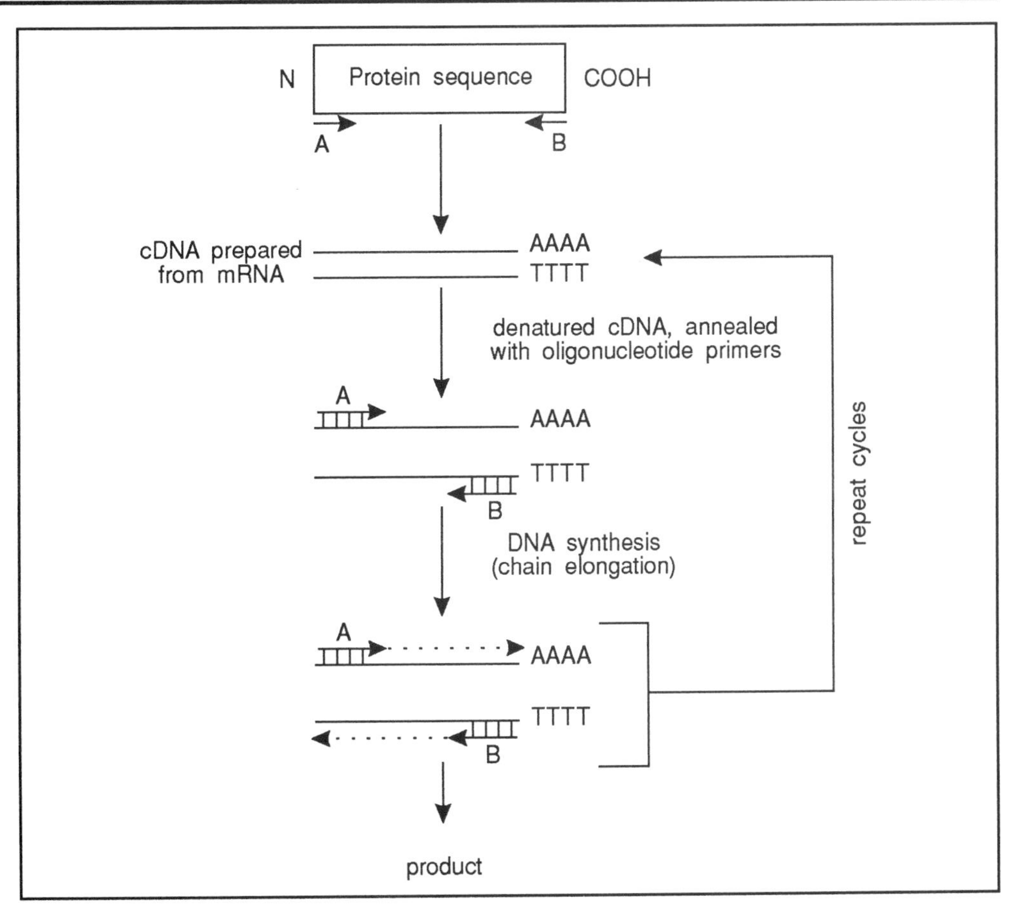

Figure 7.11 A schematic representation of the polymerase chain reaction procedure. A and B are oligonucleotides which carry the nucleotide equivalent to the N and C terminal amino acids of the protein.

7.9.2 Detection of an antibody response to the micro-organism

After infection, the host will respond with formation of antibodies to the causative micro-organism. This antibody response can be used to make a definitive diagnosis.

paired sampling technique

Two blood samples are normally taken from infected food animals, one during the acute phase of the disease and another about three weeks later during the convalescent phase. These 'paired' samples are simultaneously examined in a serological test. A fourfold or greater rise in antibody titre is considered as evidence of an infection around the first blood sampling.

Π Can you suggest a disadvantage of this procedure for serological diagnosis?

Before the results are known often four weeks have elapsed. It should be obvious that it may then be too late for the veterinary practitioner to take therapeutic or other measures.

A way of circumventing this problem is to use a serological test that detects IgM antibodies against the agent. After a primary infection, the immune system responds with the formation of antibodies of the IgM class before antibodies of the IgG class are produced. In Figure 7.12 the time course of typical IgM and IgG responses to the viral inoculation of calves is shown. Thus, around ten days after the onset of the disease a blood sample can be collected and tested in the laboratory for the presence of specific antibodies of the IgM class.

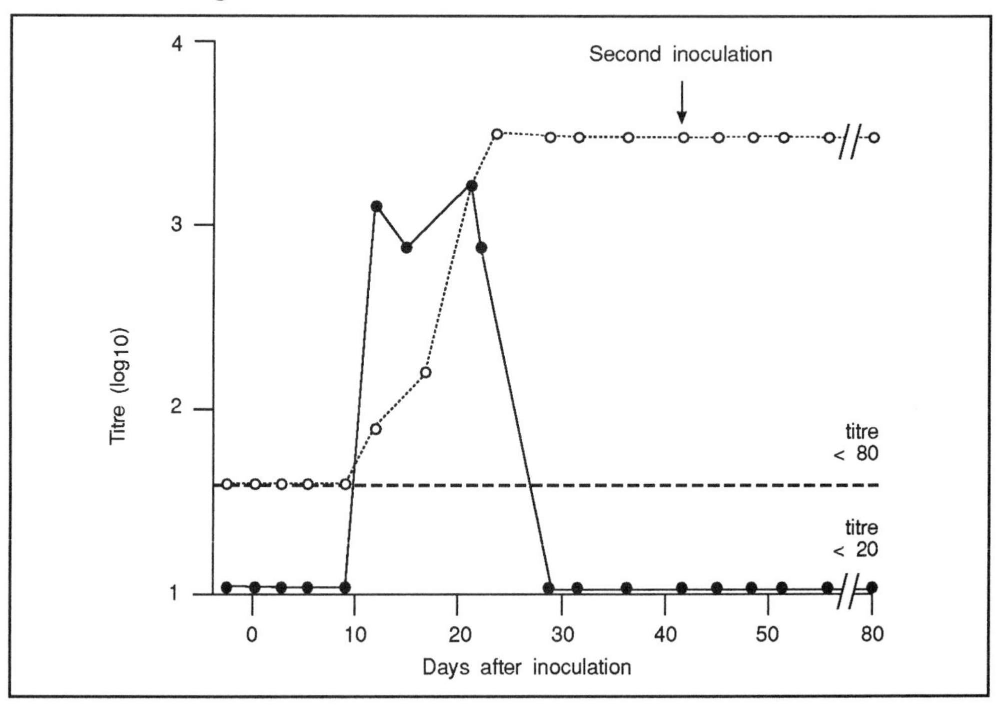

Figure 7.12 Isotype specific-BRSV (bovine respiratory syncytial virus) antibody titres in sera collected from a group-A calf, 7 months of age, which was inoculated on day 0 and reinoculated on day 42. ● —— ●IgM antibody response; ○ —— ○ IgG antibody response. (Data from Westenbruik and Kimman 1987, American Journal of Veterinary Research 48, 1132-1137).

Π List two advantages this test have over the 'paired' sample technique. (Figure 7.12 should help you).

Compared with the serodiagnosis on paired serum samples, the advantages of making a diagnosis on the basis of an IgM specific response are the reduced time necessary for diagnosis and the fact that only one sample is required.

Figure 7.13 shows the principles of an ELISA for detecting an IgM specific response to bovine respiratory syncytial virus (RSV).

An anti-IgM monoclonal antibody is immobilised on a support. The test sample is added. If it contains IgM this will bind with the immobilised anti-IgM. Then an appropriate antigen (RSV) is added. If the corresponding IgM was present in the test sample present, the antigen is bound. Another antibody against the antigen is added. This antibody is linked to an enzyme (E). If this antibody binds to the Ag, it can be detected by measuring the enzyme (E).

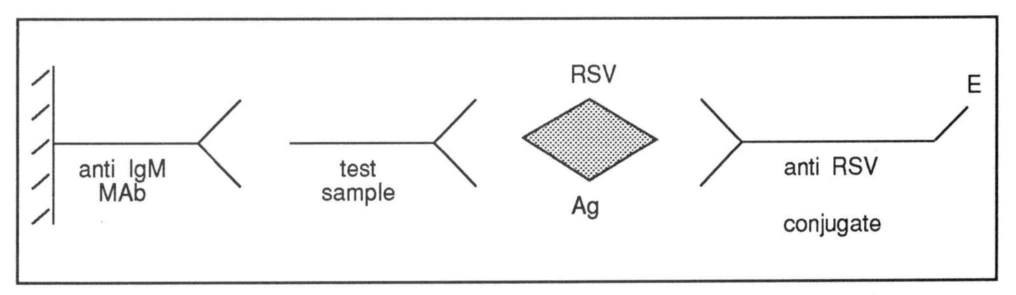

Figure 7.13 The principle of an ELISA for detection of an IgM specific response to bovine respiratory syncytial virus (see text for details).

differential diagnostic tests

An advantage of most biotech vaccines is that their antibody response can be differentiated from the antibody response after infection. This is crucial for control programmes of infectious animal diseases. The differentiation is based on measuring an antibody response against (portions of) a protein, that is (are) not expressed by the vaccine, and against which an antibody response is elicited after infection. The test may involve the use of synthetic oligopeptides or recombinant DNA polypeptides as antigen, or may be based on the use of specifically selected MAbs. At present, in the veterinary field such differentiable diagnostic kits are only available for measuring antibodies to gI, gX or gIII of Aujeszky's disease virus.

If we produce a deletion mutation vaccine for Aujeszky's disease, then such a vaccine will stimulate the production of antibodies for all of the Aujeszky's disease antigens except for the deleted antigen. Thus a deletion mutation vaccine composed of viruses in which the gX gene is deleted, will not stimulate the production of anti-gX antibodies. Wild type viruses, which contain active gX genes, will, of course, produce gX gene products and thus stimulate the production of anti-gX antibodies. Thus by detecting the presence of anti-gX antibodies will enable us to differentiate between vaccinated animals and animals that have been infected by the natural virus.

Diagnostic tests must be sensitive and specific, and must give reproducible results. The sensitivity of tests for detecting pigs infected with Aujeszky's disease virus in vaccinated populations can best be evaluated by detecting antibodies in sera of pigs that are first vaccinated and then challenge-infected with a virulent virus. These antibodies must persist for a long period after infection. The evaluation of the test's specificity should be carried out with sera from repeatedly vaccinated animals and from animals infected with unrelated organisms.

7.10 Conclusions

Biotechnology has revolutionised research on the development of vaccines resulting in new insights into the molecular mechanisms of pathogenesis and the immunology of infectious diseases. In the late 18th century, Jenner protected humans against smallpox by vaccinating them with a cowpox virus, and some 100 years later Pasteur described attenuation of micro-organisms.

Almost all vaccines currently used are produced by conventional means. There are live and killed 'conventional' vaccines. Live vaccines contain a replicating agent, whereas killed vaccines contain a noninfectious agent or a subunit thereof and an adjuvant. In general, live vaccines stimulate the strongest protective immunity, but may present hazards, among which are contamination with adventitious agents, whereas killed vaccines are relatively poor immunogens, but are usually safer.

The efficacy of vaccines is examined in the target animal. The animals are vaccinated and challenge-infected and are monitored for disease in comparison with unvaccinated animals. Most safety tests are also conducted in the target animal. A good vaccine is efficacious and causes no harm.

Biotechnology has offered new insights into the molecular mechanisms of pathogenesis and immunology, and has opened perspectives for modern vaccines, such as subunit recombinant DNA vaccines, chemically synthesised peptide vaccines, anti-idiotype vaccines, deletion mutant vaccines and vector vaccines. A handful of these biotech vaccines are on the market. These modern approaches may pave the way for vaccines against diseases against which a vaccine is not yet available, eg parasitic vaccines.

The laboratory diagnosis of infectious diseases is based on detection of the microbe, its antigens or part of its genome, and on detection of an antibody response against the microbe. The polymerase chain reaction is a new and highly sensitive technique to detect microbial DNA/RNA. IgM ELISAs have been developed for rapid serodiagnosis, and viral protein specific ELISAs for differentiating between infected and vaccinated animals.

Summary and objectives

This chapter included a brief history of vaccinology, an introduction to the immune system, and an overview of general principles of vaccines produced by conventional means. The modern approaches to vaccine development were described as well as the biotech vaccines that are already on the market. Finally, a section on modern diagnostic methods completed the chapter.

After having studied this chapter you should be able to:

- summarise the history of vaccinology;
- differentiate between systems and mucosal immunity;
- describe the nature of conventional vaccines and evaluate the advantages and disadvantages of live and killed vaccines;
- describe animal experiments conducted to evaluate the efficacy and safety of vaccines;
- appreciate the immunogenicity of microbial proteins and its importance for developing recombinant DNA subunit vaccines;
- list the various approaches used in the development of biotech vaccines, and explain the differences between these approaches;
- use the criteria of safety and efficacy to select strategies for producing vaccines;
- describe some of the principles and applications of modern diagnostic methods.

Responses to SAQs

Responses to Chapter 1 SAQS

1.1

1) GMMO = genetically modified micro-organisms.

2) GMO = genetically modified organism.

3) Mutagenesis, self-cloning of some non-pathogenic organisms (see Appendix 1, Section 1.4).

4) Type B operations are large scale (10 litres or greater) operations used for commercial purposes (Appendix 1, Section 1.5).

5) Group II - *Salmonella typhi* is a pathogen and does not fall into group I (see Appendix 1, Section 1.5).

6) To the competent authority of the Member State within whose territory the release is to take place (Appendix 1, Section 2.4). This notification should contain information set out in 5 headings:

   - general information;
   - information relating to the GMO;
   - information relating to the conditions of release and the receiving environments;
   - information relating to the GMO and the environment;
   - information on monitoring, control, waste treatment and emergency response plans (Appendix 1, Section 2.4).

Responses to Chapter 2 SAQs

2.1

1) Releasing and inhibiting factors (or hormones) from the hypothalamus which affect the adenohypophysis, for example GnRH and PIF.

2) Gonadotrophins from the adenohypopysis which stimulate the functions of the gonads.

3) The sex hormones or steroids which affect the genital tract.

2.2

1) We would anticipate a period between August and January would be the most likely oestrous season. Our reasoning is that the young would be borne as food becomes most available (spring/early summer). At latitude 50-55° this is a period February-July. Thus fertilisation, which takes place 18 months previous to the birth of the young, must take place in August-January. Thus, this is the most likely period to be the oestrous season.

2) The most likely link is day-length although temperature might also affect the onset of the oestrous season.

2.3

Progesterone is the main regulator of cyclicity within oestrus. It exerts a constant negative feedback effect on tonic LH release, but changing levels of the steroid vary the force of this feedback on a regular cycle to influence LH levels.

Oestradiol-17β regulates the rhythm of oestrous and anoestrous seasons. Its feedback effects vary from negative to positive with increasing and decreasing daylength respectively. Hence as photoperiod increases negative feedback depresses LH release and the animal enters the anoestrous season, but as photoperiod decreases progressively increased positive feedback, LH and oestradiol release bring on the oestrous season.

2.4

Oxytocin is conveyed directly by neurosecretion from cell bodies in the hypothalamus to the nerve endings in the neurohypopysis for release into blood capillaries. However, GnRH is carried via portal blood vessels in the median eminence from hypothalamic neurons to the adenohypophyseal cells that synthesise gonadotrophic hormones.

2.5

1)	Luteinising hormone	steroid synthesis, oocyte maturation, ovulation, support corpus luteum
2)	Follicle stimulating hormone	oocyte maturation, steroid synthesis
3)	Prolactin	support corpus luteum, lactation

(Note the overlap of some functions)

2.6

All of the statements (1-11) are true. If you get them all right you have certainly understood the events of the oestrous cycle.

2.7

A silent heat denotes the first ovulation of the oestrous season that occurs without oestrous behaviour due to the absence of pre-exposure to progesterone. Nevertheless it may serve to signal the imminence of oestrus in ewes to the rams. A silent heat also leads to the formation of a corpus luteum which will secrete progesterone. The progesterone influences the genital tract to ensure a proper micro-environment in the tract for gamete transport and development of a fertilised ovum at the subsequent oestrus.

Responses to Chapter 3 SAQs

3.1

They both depend on manipulation of the time of regression of the corpus luteum. Prostaglandins are used to cause regression of the functional corpus luteum, and two PG injections effectively serve the same function in animals in early luteal, or follicular phases. Progestagen treatment mimics the negative feedback inhibition of gonadrophin hormone secretion by progesterone from the corpus luteum, leading to oestrus and ovulation once the inhibition is removed.

3.2

No - it would not be possible to implant an embryo into the recipient animal at any stage of its oestrous cycle.

There needs to be good synchronisation between donor and recipient for successful implantation. Acceptable pregnancy rates will be obtained only if the stage of the oestrous cycle in the recipient is within one day of that of the donor at the time of collecting the embryos. The cycles of recipients of embryos that have been frozen previously must also match those of the donors at embryo recovery.

3.3

Detection of high levels of progesterone in bovine milk following insemination would certainly be indicative of pregnancy. Time of testing is very important, and the occurrence of embryo loss later than the time of testing could lead to false positive pregnancy diagnoses. However, non-pregnant animals have low progesterone concentrations at the time of sampling and can be detected with over 95% accuracy. Hence the test is valuable to check that the pregnancy is still present, not having spontaneously aborted (see second arrow on the time scale of Figure 3.5).

3.4

1) False. Synchronisation of oestrus and ovulation by holding animals in the luteal phase is achieved by administering progesterone (or progestagens) not prostaglandins (eg PGF2α). PGF2α causes repression of the corpus luteum (ie it terminates the luteal phase).

2) True.

3) True.

4) False. LH levels do not characteristically rise or fall during pregnancy and thus measuring LH levels would not be a good method of detecting pregnancy.

5) True. Prostaglandins cause breakdown of the corpus luteum. Thus the levels of progesterone falls. High levels of progesterone are needed for a successful pregnancy.

Responses to Chapter 4 SAQs

4.1

Embryo-splitting and nuclear transfer techniques yield sets of embryos, each consisting of several individual embryos with identical genotypes. In embryo splitting all cells taken from a single embryo contain identical genetic material, likewise the nuclei transplanted from a single embryo into enucleated oocytes also contains the same genetic material. However embryos derived from superovulation or *in vitro* maturation of oocytes and fertilised with a single batch of sperm share common genetic material from their 'parents', but not identical genotypes.

4.2

Technique	Function
superovulation	to produce multiple oocytes
enucleation	to provide cells into which blastomere nuclei can be transferred
microfluorescence enzymology	to detect metabolic status of embryos
electrofusion	to fuse blastomeres and oocytes
DNA probing	to enable sexing of embryos

You should of course realise that many of the techniques cited in this SAQ also have applications in many other areas of biological study.

Responses to Chapter 5 SAQs

5.1 Even if the chimaeric animals so produced proved fertile there is no guarantee that the foreign DNA will be inherited by their offspring (let alone be expressed). For the DNA to pass to the next generation it would have to have been added to the cells of the embryo that contribute to the germ line and so eventually to the gametes. Thus chimaeras may, or may not, pass on the foreign DNA, dependent on whether the gonadal tissues include the introduced genes.

5.2 You had some scope in responding to this question. The important points are as follows. Transgenic animals are created by the addition of new DNA into the germ line, where the new genes can be inherited in a Mendelian manner. However the expression of the genes in the progeny may be variable due to several factors. The key to producing true transgenic animals is the incorporation of the foreign DNA into cell lines which eventually provide the gametes. Successful micro-injection of the pronucleus of the zygote means that the new DNA will be in all the cells of developing embryos, including those of the germ line. Use of retroviral vectors with multicelled embryos will only produce transgenic animals when germ line cells are 'injected', and chimaeric offspring will often result. Similarly with the incorporation of transfected ES cells into early embryos the offspring will be chimaeric, with only a possible inclusion of the new DNA in mature gamete producing tissues, there is no guarantee of the new genes being inherited by the offspring. However the ES system does have many potentially exciting applications for the introduction of transgenes.

5.3 The sequence we would have written is:

- isolate $mRNA_x$ from cat;
- make a cDNA copy of $mRNA_x$;
- attach a poly A tail to the $cDNA_x$;
- use a cDNA against a mouse liver specific gene product to enable the identification and isolation of a mouse liver specific promoter (LS-promoter);
- attach the LS-promoter to a $cDNA_x$-poly A tail construct;
- insert an LS-promoter-$cDNA_x$-poly A tail construct into a bacterial vector;
- infect a bacterial culture with a vector carrying an LS-promoter-$cDNA_x$-poly A tail construct;
- isolate multiple copies of the construct from the bacteria;
- inject LS-promoter-$cDNA_x$-poly A tail construct into mouse pronuclei.

You should realise that for simplicity, we have omitted many of the details of each stage (eg the screening needed to identify the appropriate mRNA, how poly A tails are isolated etc). These techniques are described in more detail in the BIOTOL text 'Techniques for Engineering Genes'.

Responses to Chapter 6 SAQs

6.1 Bioassay relies on the measurement of an actual physiological function of the hormone and so reflects directly the biological potency of a purified hormone extract or synthetic hormone preparation. Radioreceptor assays use the property of specific hormone receptor proteins to bind competitively molecules of the hormone and its radiolabelled counterpart. So in effect this technique can measure the amount of pure, biologically active hormone present. Radioimmunoassay works in a similar way but uses the binding properties of a specific antibody raised to the pure hormone. Hence the latter technique may to some extent reflect the immunogenic properties of the hormone molecule rather than just its biological activity. However all three techniques are very valuable in hormone assays.

6.2 The appropriate words are:

1) homeorhetic;

2) increases;

3) antagonises;

4) hyperglycaemia;

5) enhances;

6) stimulate.

If you could not answer this SAQ, read sections 6.3.2 and 6.3.3 again.

6.3 It should be clear that administration of GHRH should increase production and release of ST into the bloodstream. This method is attractive in that the animal is then producing its own ST, possibly releasing the hormone in several biologically active forms. Also it seems more effective to manipulate a hormone's control mechanism with minute quantities of exogenous material than to dose somatotropin itself - in much larger quantities. Examples of a stimulated milk yield in dairy cows treated with exogenous GHRH have been documented.

6.4

1) False. There is no reason to believe that changing the promoter would alter the state of oxidation or reduction of the translated gene product.

2) False. Similar reasoning to 1).

3) False. The gene β-galactosidase in *E. coli* is repressed by glucose (catabolite repression). Thus growing an *E. coli* containing a bST construct under the control of a β-galactosidase promoter in a glucose medium would mean that the expression of the construct would be repressed (ie switched off). Yields of bST would, therefore be very small.

6.5 Although the bovine mammary tissue does not contain bST receptors it does contain receptors for IGF-I and II. Since both hormones are found elevated in the circulation after bST treatment, the effect of bST on the udder could therefore be mediated by the IGFs. Indeed, infusion of IGF-I directly into one mammary gland of lactating goats resulted in an increase in the rate of milk secretion by the infused gland.

The positive milk response after bST injection in the cow is rapid, just as is the fall in milk yield after cessation of bST treatment. It is therefore likely that the indirect effect of bST is produced via an increase in the secretion rate of mammary epithelial cells rather than a change in mammary cell numbers.

6.6 These two polypeptides have a short biological half-lives and denature during pasteurisation, which is common practice in dairy processing. Oral intake of bST is without risk, because it is digested like any other natural protein. bST and pST are species specific and they cannot react with human ST receptors. These factors mean that meat from ST-treated animals is also safe for the consumer. Perhaps the greater chance of risk comes from increased IGF-I levels. Human, bovine and porcine IGF-I have identical structures and porcine IGF could have an effect on the human system. There is however little evidence of ingested hormone ever reaching its receptors. Thus it is unlikely to have any affect.

Responses to Chapter 7 SAQs

7.1 For infections with micro-organisms that remain restricted to the mucosae it appears logical to apply a local or 'mucosal' vaccination to stimulate IgA production for a protective immunity at the mucosae. Parental (usually intramuscular) vaccination induces mainly a systemic immunity, and is often sufficient to prevent a systemic infectious disease, which is frequently characteristic of a transient viraemic phase.

7.2 Attenuation may be considered a primitive form of genetic engineering, because it is aimed at inactivating, or weakening, genes that are involved in virulence, by inducing mutations, shifts, replacement or deletions, delet, of genes. It is primitive, because the virulence genes were not known, and because attenuation, mutation of virulence genes was achieved merely by chance instead of by a well-defined strategy.

7.3 Theoretically, a live vaccine will best meet the efficacy requirements for an ideal vaccine, because it replicates in the host, giving rise to a wide range of immune responses, and thus simulates a natural infection better than a nonreplicating vaccine. It tends to survive longer in the host and thus stimulation is prolonged.

7.4

1) Efficacy	**Live**	**Killed**
a - broad immunity	+	
b - durable immunity	+	
c - mucosal administration	+	
d - number of vaccinations	+	
2) Safety		
a - residual virulence		+
b - persistence		+
c - reversion to virulence		+
d - transmission		+
e - recombination with wild-type virus		+
f - contamination with extraneous agents		+
g - local and systemic side-effects		+
h - incomplete inactivation		+
3) Cost		
	variable	

7.5 Vaccines derived from human blood may be contaminated with adventitious agents, for instance human immunodeficiency virus, the causative agent of AIDS. Such vaccines must be highly purified to be safe and are very expensive. There is also a limited supply of antigen available.

7.6 The five approaches are:

1) recombinant DNA subunit vaccines, produced in bacteria, yeast etc;

2) chemically synthesised peptide vaccines;

3) anti-idiotype vaccines;

4) deletion mutant vaccines;

5) vector vaccines.

Reassortant vaccines do not strictly require biotechnology in their development.

7.7

1) A - Adjuvants may be added to attenuated viruses but they are not usually necessary. With inactivated viruses and subunit (purified antigens) vaccines, adjuvants are usually necessary to stimulate a greater immune response.

2) B and C - This feature tends to favour inactivated (killed) viruses or subunit vaccines. Neither will revert into a virulent form.

3) A - Live vaccines stimulate a greater immune response than inactive ones.

4) A and especially B. If knowledge of the specific antigens produced by a virus is lacking, it is difficult to imagine how such antigens can be selected to produce a subunit (purified antigen) vaccine. You should remember that most viral diseases can be caused by a 'family' of slightly different viruses each with a slightly different collection of antigens. It is important to choose an antigen present on all forms of the viruses (ie a common epitope) to provide full protection.

5) B and C. Subunit and inactivated viruses are more stable than live, attenuated viral suspensions.

6) C - Purified antigens are chemically and physically simpler than complete viral particles. It is, therefore, easier to characterise them more thoroughly.

7.8 We will discuss the merits or otherwise of each of the three strategies in turn.

Strategy 1 - This is a strategy for producing a subunit (M1) vaccine. It is, however, unlikely to work. Although M1 is produced by all of the viruses tested it does not seem to be part of the viruses. Thus although a vaccine using M1 would stimulate antibodies against M1, these antibodies would not inactivate this molecule as it is produced and used within the host cells.

Strategy 2 - This has many similarities to strategy 1 (ie a subunit vaccine) except that capsomer proteins are to be used. Because of the epitopic differences between the capsomers, this strategy is also unlikely to work.

Strategy 3 - This looks the most promising of the three strategies. It aims to produce an attenuated virus vaccine. From the information given in the question, we could visualise producing a deletion mutation using *Eco* R1 in which a reconstructed viral genome, without the M1 gene could be produced. Thus:

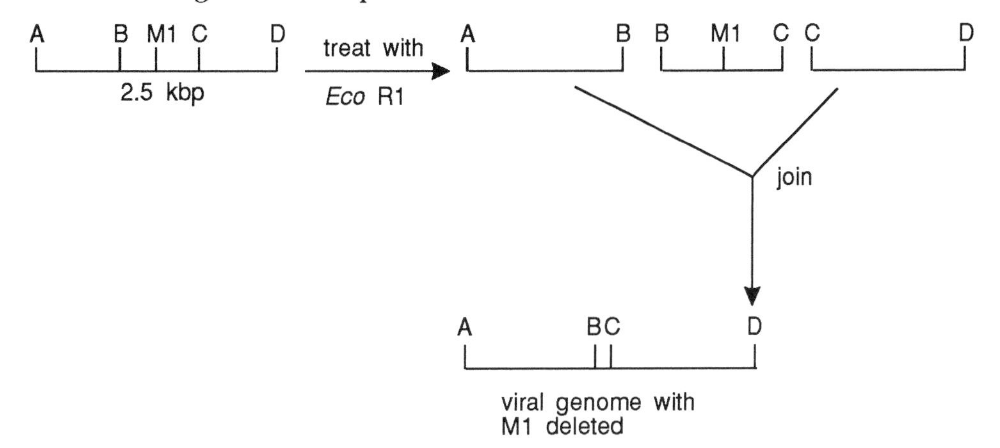

Such a virus would have many antigenic determinants and might offer universal protection against the viruses causing the disease. Since M1 is needed for maturation, the strain would almost certainly be attenuated. Whether or not it would be attenuated too much or too little would require thorough testing.

Responses to Appendix 2 SAQs

1

1) False. It is produced by the hypothalamus.

2) True (see Figure 3).

3) False. TSH release in inhibited by thyroxine, not by testosterone.

4) True (see Figure 2).

2

The correct order is 5, 6, 2, 3, 4, 1.

3

1) True.

2) False. Steroid hormones are synthesised as required; they are not stored in the cell.

3) True.

4) False. Steroid and thyroid hormones do not require calcium ions for their release.

5) True.

4

No. 3 is the correct answer. If a cell with receptors for a given hormone is exposed to a high plasma level of that hormone for a prolonged time, there will be a reduction in the number of receptors on the cell.

5

The correct order is 2, 6, 4, 5, 7, 3, 1

6

No, the response would not be completely abolished because the intracellular calcium level could still be increased by calcium ions being released from intracellular stores. This would be mediated by inositol triphosphate.

7

1) False. IP_3 acts to increase the cytosolic concentration of calcium ions by releasing them from intracellular stores.

2) True.

3) False. Phosphodiesterase inactivates cAMP only.

4) False. The different transduction mechanisms are mediated by different G-proteins.

5) True.

8

A	cAMP	Maximum responses are achieved when adenylate cyclase is activated by forskolin and when theophylline is added. Both these would result in an increase in the concentration of cAMP.
B	Calcium ions	The ionophore, when given, results in a maximum response ie. when the concentration of calcium ions is increased.
C	cAMP+ calcium ions	Only 25% of the maximum response is achieved when agents altering either cAMP or calcium ion levels are given individually. When theophylline and the ionophore are given together, a full response is seen, indicating that both second messenger systems act synergistically to give the full response in this case.

9 Your drawing should have looked like this:

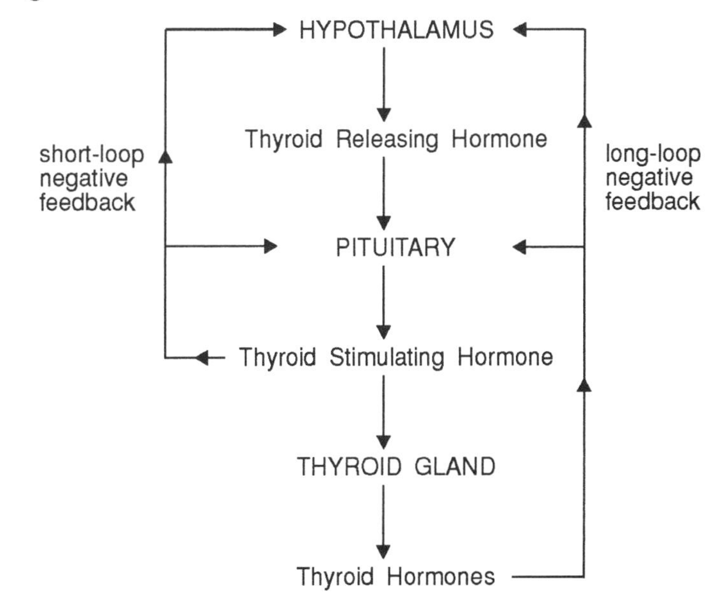

10

Hormone	Function
TRF	thyroid hormone secretion
oxytocin	smooth muscle contraction
testosterone	spermatogenesis
erythropoietin	red blood cell production
gastrin	hydrochloric acid release
aldosterone	sodium reabsorption

11

Hormone	Site of production
insulin	pancreas β cells
calcitonin	thyroid
erythropoietin	kidneys
cortisol	adrenal cortex
adrenaline	adrenal medulla
TRF	hypothalamus
luteinising hormone	anterior pituitary
glucagon	pancreas α cells
progesterone	corpus luteum
ANF	cardiac atria

Appendix 1 - EC Directives

1 EC directive on the contained use of genetically modified micro-organisms

The EC directive on the contained use of genetically modified micro-organisms is based on the OECD guidelines and is published in the Official Journal of the European Communities, L117, Volume 33, 8 May 1990.

1.1 Purpose of the directive

Article 1 gives the purpose of the directive as:

"to lay down common measures for the contained use of genetically modified micro-organisms with a view to protecting human health and the environment".

1.2 Scope

The scope of every regulation depends on the definitions and the exemptions.

This directive covers contained use of genetically modified micro-organisms. In the following paragraphs, therefore definitions of the terms genetically modified micro-organisms and contained use are given, and exemptions are explained.

1.3 Definitions

1.3.1 Micro-organisms

Micro-organisms are defined as:

"any microbiological entity, cellular or non-cellular, capable of replication or transferring genetic material".

It should be noted that this directive only covers micro-organisms, and not all organisms. It was argued that including all organisms in this directive would give an unacceptable delay in the implementation of the proposals. The Commission undertook to keep the whole biotechnology sector under review and make appropriate proposals to extend the scope of this directive to genetically modified organisms. The national Member States may maintain and adopt national measures for the contained use of organisms other than micro-organisms.

Genetically modified micro-organisms are in this document abbreviated as GMMOs.

1.3.2 Genetically modified micro-organisms

Genetically modified micro-organisms are defined as:

"micro-organisms in which the genetic material has been altered in a way that does not occur naturally by mating or by recombination".

In order to specify this, a list of techniques is given by which genetic modification can occur.

This non-limitative list is Annex Ia, of the Directive and contains:

- recombinant DNA techniques using vector systems;
- techniques involving the direct introduction of heritable material prepared outside the micro-organism, such as micro-injection;
- cell fusion and hybridisation techniques by means or methods that do not occur naturally.

It should be noted here that the scope of the regelation is not limited to recombinant DNA techniques. This gives recognition to the fact that a variety of molecular genetic transformation techniques, including recombinant DNA, are widely used and may have similar safety considerations.

1.3.3 Contained use

Contained use means:

"any operation in which micro-organisms are genetically modified or in which such organisms are cultured, stored, used, transported, destroyed or disposed of and for which physical barriers together with chemical and/or biological barriers, are used to limit their contact with the general population and the environment".

The key term here is the physical barrier.

1.4 Exemptions

The directive does not apply where genetic modification is obtained through the use of certain techniques (Article 3). These techniques are:

- mutagenesis;
- construction and use of somatic animal hybridoma cells;
- cell fusion of plants which can also be produced by traditional breeding methods;
- self cloning of certain non-pathogenic naturally occurring micro-organisms.

The directive does furthermore not apply to:

- the transport of GMMOs;

- GMMOs which have been placed on the market under Community legislation (Article 5).

1.5 System of the directive

1.5.1 Group I and Group II organisms/Type A Type B operations

Group I and Group II organisms

For the purpose of the directive, micro-organisms are classified in two groups (Article 4).

Group I - those satisfying certain criteria.

Group II - those other than group I.

Group I organisms have a long record of safe use and are considered to be safe when used under specific conditions. The criteria for Group I are given in Annex II of the directive, which gives criteria for the recipient and parental organisms (non pathogenic etc), to the vector and the insert used, and to the final GMMO. This Annex is based on the criteria for GILSP (Good Industrial Large Scale Practice) set up by the OECD.

For GMMOs of Group I, principles for good microbiological practice and of good occupational safety and hygiene shall apply. These principles are also based on the OECD report of 1986.

In addition to these principles, certain containment measures set out in an Annex shall be applied to ensure a high level of safety for GMMOs of Group II (Article 6).

Type A and Type B operations

For the purpose of the directive, a distinction is made between Type A and Type B operations.

Type A operations are operations used for teaching, research and development, or non-industrial or non-commercial purposes and which are of a small scale (eg 10 litres volume or less).

Type B operations are operations other than operations of type A.

1.5.2 System of the directive

The regulatory system of this directive contains two sorts of procedures:

- activities for which a notification is required;

- activities for which an authorisation is required.

The two distinctions can be combined to four possible activities:

- Type A operations with Group I - IA operations;

- Type B operations with Group I - IB operations;

- Type A operations with Group II - IIA operations;
- Type B operations with Group II- IIB operations.

In addition to this, the first use of an installation for an operation involving GMMOs is considered to be an activity for which a procedure is required.

Articles 6, 7 and 8 of the Directive assign specific procedures to each of the possible activities mentioned above and form the basis of the procedures.

The possible procedures are:

- to keep records of the work carried out and make them available to the competent authority (Article 9.1);
- a notification within a reasonable period before commencing the use (Article 8);
- a notification and a waiting period (Article 9.2 and 10.1);
- an authorisation (Article 10.2).

The possibilities are presented below

Type of operation	Procedure	Article
First use in an installation	notification within reasonable period in advance	8
IA operations	keep records	9.1
IB operations	notification and waiting period of 60 days plus additional 60 days at request of competent authority	9.2
IIA operations	notification and waiting period of 60 days plus additional 60 days at request of competent authority	10.1

1.6 Additional provisions

Articles 11 to 14 lay down some specific obligations for the Member States:

- to ensure that, where necessary, before an operation commences an emergency plan is drawn up and information on safety measures is available (Article 14);
- to ensure that, in case of an accident, the proper measures and steps will be taken (Article 15);
- consult, when necessary, with other Member States and inform the Commission (Article 16);

- to ensure that inspections are carried out (Article 17a).

1.7 Confidentiality

Article 19a:

"The Commission and the competent authorities shall not divulge to third parties, any confidential information notified or exchanged under this directive and shall protect the intellectual property rights relating to the data received".

The notifier indicates what information needs to be kept confidential. However, it is the competent authority which decides, after consultation with the notifier, which information shall be kept confidential.

2 EC Directive on the deliberate release of genetically modified micro-organisms

This directive of 23rd April 1990 is published in the Official Journal of the European Communities, L117, Vol 33, 8 May 1990.

2.1 Parts of the Directive

This directive consists of four parts:

Part A - General provisions;

Part B - Research and Development (R & D) and introductions into the environment other than placing on the market;

Part C - Placing products on market ;

Part D - Final provisions.

2.2 Part A: General provisions

2.2.1 Purpose of the Directive

The purpose of this directive is laid down in Article 1:

"to approximate the laws, regulations and administrative provisions of the Member State and to protect human health and the environment when carrying out a deliberate release or placing on the market of genetically modified micro-organisms".

In order to gain a better understanding of the purpose and background of this directive, the considerations in the preamble should also be studied.

In addition to the purpose of this directive Article 4 emphasises in general terms the obligations of Member States in accomplishing this purpose.

2.2.2 Scope of the Directive-definitions

Repeating what was explained earlier: the scope of every regulation depends on the definitions and exemptions.

This directive covers the deliberate release of genetically modified organisms. In the following paragraphs the definitions of the directive are explained. These definitions are summed up in Article 2.

Organism

Organism is defined in this directive as:

"any biological entity, capable of replication or of transferring genetic material".

Since the term 'biological entity' is open for multiple interpretation, an explanation is given in the statements for inclusion in the Council's minutes:

"This definition covers: micro-organisms, including viruses and viroids; plants and animals; including ova, seeds, pollen, cell cultures and tissue cultures from plants and animals".

Hereafter, the term genetically modified organisms is abbreviated to GMO.

Genetically modified organism

The definition of a genetically modified organism is analogous to the definition of a genetically modified micro-organisms, provided that the term 'micro-organism' is replaced by the term 'organism'.

Deliberate release

Deliberate release is defined in paragraph 3 of Article 2 as:

"any intentional introduction into the environment of a GMO or a combination of GMOs without provisions for containment such as physical barriers or a combination of physical barriers together with chemical and/or biological barriers used to limit their contact with the general population and the environment".

A further clarification is given in the Council's Statements:

"the introduction by whatever means, directly or indirectly, by using, storing, disposing, or making available to a third party".

By using the terms "without provisions for containment such as..." as a cross reference to the directive for contained use, a complementary system is achieved.

In other words: every activity that is not a contained use is regarded as a deliberate release.

2.2.3 Exemptions

The exemptions of the scope of this directive are laid down in Article 3:

"This directive shall not apply to organisms obtained through the techniques of genetic modification listed in Annex Ib". These include:

- mutagenesis;
- cell fusion of plant cells when the plant can also be produced by traditional methods.

The background of these exemptions is that these specific applications have been used in a number of applications and have a long safety record.

2.3 System

The system of this directive is based on two notions:

- the release of a GMO into the environment can have adverse effects on the environment which may be irreversible;
- GMOs, as well as other organisms, are not stopped by national frontiers.

These two notions led to the choice of a system whereby:

- every introduction of a GMO into the environment is subject to an authorisation by the competent authority of the country where the introduction takes place;
- before an authorisation is given, the competent authority consults the other Member States of the Community.

In addition to this, a distinction between Research and Development (R & D) and placing on the market is made.

2.3.1 R & D and placing on the market

In this directive, a distinction is made between:

- Research and Development (R & D) and other introductions into the environment than placing on the market (part B of the directive) ;
- Placing on the market (part C of the directive).

The result of this distinction and the system of the directive is that placing on the market involves a system of international consultation whereby the competent authority cannot take a decision without the agreement of the other Member States.

All other introductions into the environment (Part B, which basically consists of R & D introductions) are subject to an authorisation of the competent authority which may give its decision without the approval of other Member States, though be it that these introductions are also notified to the other Member States who may give comments.

The reason for this distinction, is found in the numbers and the spread of a GMO connected with placing on the market. When a product containing GMOs or consisting of a GMO is placed on the market, it will be spread all over Europe in, possibly, vast numbers under uncontrolled circumstances. Whereas R & D introductions are normally small scale introductions of a limited number of GMOs and under controlled circumstances.

2.4 Part B: Research and Development (R & D) and introductions into the environment other than placing on the market

The basis of part B of this Directive is laid down in the combination of the Articles 5, paragraph 1 and Article 6, paragraph 4.

Article 5, paragraph 1 states:

"Any person before undertaking a deliberate release of a GMO for the purpose of research and development or for any other purpose than placing on the market, must submit a notification to the competent authority of the Member State within whose territory the release is to take place".

Article 6, paragraph 4, states:

"The notifier may proceed with the release only when he has received the written consent of the competent authority, in conformity with any conditions required in this consent".

2.4.1 The notification

Article 5, paragraph 2, gives the requirements of a notification under part B of the directive:

"The notification shall include the information specified in Annex II".

Annex II is an indicative list of points of information set out under 5 headings:

I General information;

II Information related to the GMO;

III Information relating to the conditions of release and the receiving environment;

IV Information relating to the GMO and the environment;

V Information on monitoring, control, waste treatment and emergency response plans.

This Annex II is based on the OECD report of 1986. It is essential to realise that this Annex contains an indicative list, and that not all the points included will apply to every case.

2.4.2 Authorisation

The authorisation procedure is laid down in Article 6.

Paragraph 1: "On receipt and after acknowledgement of the notification the competent authority shall examine the conformity of the notification with the requirements of this directive".

Paragraph 2: "The competent authority, having considered where appropriate, any comments by other Member States, shall respond in writing to the notifier within 90 days by indicating either:

- that the release may proceed;
- that the release does not fulfil the conditions of this directive and the notification is therefore rejected.

For calculating the waiting period of 90 days, the period needed for the notifier to supply further information and the period in which a public inquiry is carried out, shall not be taken into account.

The third paragraph of Article 6 gives the steps to be taken when new information becomes available with regard to the risk of the product. In that case the notifier shall revise the information, inform the competent authority and take the necessary measures to protect human health and the environment.

2.4.3 International consultation

Within 30 days after the receipt of a notification, the competent authority shall send to the Commission a summary of the notification. The Commission shall immediately forward these summaries to the other Member States which may, within 30 days, present observations. It should be stressed here that these observations are not binding to the original competent authority.

2.5 Part C: Placing on the market products containing genetically modified organisms

2.5.1 General provisions

Part C of this directive stars with Article 10, which gives in paragraph 1, a set of general conditions before any product can be placed on the market.

These conditions are that:

- consent has been given under part B of the directive, meaning that no GMO can be placed on the market without a proper R & D stage;
- the product should comply with this directive and relevant product legislation.

Paragraph 2 of Article 10 indicates that the procedure for placing a product on the market shall not apply to products covered by Community legislation which includes a specific environmental risk assessment similar to that provided in this directive. The background of this provision is that:

- it is desirable to have only one procedure for placing products on the market;

- product legislation already contains procedures for placing on the market.

2.5.2 System

The same system of part B is found in part C.

Article 11, paragraph 1:

"before a GMO or a combination of GMOs are placed on the market as or in a product, the manufacturer or the importer to the Community shall submit a notification to the competent authority of the Member State where they are placed on the market for the first time".

Article 11, paragraph 5:

"the notifier may only proceed when he has received a written consent".

2.5.3 The notification

The first paragraph of Article 11 says that the notification shall include the information of Annex II information obtained from R & D releases and specific product information laid down in Annex III (use, labelling, packaging etc).

The final paragraph of Article 11 gives the steps to be taken when new information becomes available with regard to the risk of the product, analogous to Article 6.

2.5.4 Authorisation

The authorisation procedure of placing on the market is in fact a two step procedure. The first step is given by Article 12:

Paragraph 1: "On receipt and after acknowledgement of the notification the competent authority shall examine the conformity of the notification with the requirements of this directive".

Paragraph 2: "The competent authority shall respond in, within 90 days, by either:

- forwarding the dossier to the Commission with a favourable opinion;
- informing the notifier that the release does not fulfil the conditions of this directive and the notification is therefore rejected;

For calculating the waiting period of 90 days the period needed for the notifier to supply further information shall not be taken into account.

2.5.5 International consultation

The second step of the procedure is laid down in Article 13:

"The Commission shall immediately forward the dossier to the other Member States which may, within 60 days, present observations that are received from the other Member States".

When an objection is received and the competent authorities concerned cannot reach an agreement within these 60 days, the commission shall take a decision in accordance to a specific procedure.

2.5.6 Placing on the market: Community wide

One of the key articles of this part C is Article 15:

"A Member State may not restrict or impede, on grounds relating to the notification and written consent of a release under this directive, the placing on the market of product containing or consisting of GMOs which comply with the requirements of this directive".

This means that when a product has received a consent after the procedure of part C, no Member State may restrict the placing on the market on grounds of protecting human health or the environment.

When a Member State has justifiable reasons (eg new information) that such a product constitutes a risk, it may provisionally restrict the product, after which the Commission shall take a decision in accordance to a specific procedure (Article 16).

The commission shall publish a list of products which received consent under this directive (Article 17).

2.6 Part D: Final provisions

2.6.1 Confidentiality

Article 19:

"The Commission and the competent authorities shall not divulge to third parties any confidential information notified or exchanged under this directive and shall protect the intellectual property rights relating to the data received".

The notifier indicates what information he wants to be kept confidential, though be it that certain information cannot be kept confidential, like the name and address of the notifier and a description of the GMO, methods for monitoring and the evaluation of foreseeable effects.

It is the competent authority which decides, after consultation with the notifier, which information shall be kept confidential.

2.6.2 Commission procedure

The specific procedure mentioned before is explained in Article 21, which in general terms says that the Commission will be assisted by a Committee which votes by qualified majority. If measures envisaged by the Commission are not in accordance with the opinion of the committee, it will be submitted to the Council.

Appendix 2 - The fundamentals of the endocrine system

1 Introduction

Much of this book is concerned with the manipulation of reproduction and growth in animal systems. Most frequently this involves some perturbation being applied to the regulatory mechanisms involved in these processes. A major part of the control of metabolism and development is operated through the endocrine system. It will be no surprise, therefore, that this text leans heavily on a knowledge of the endocrine system. The core text has been written on this assumption. Many readers, however, may have only a sketchy impression of how the endocrine system works and of the nature of the factors (hormones) involved in the regulation of metabolism. Others, may have need to be reminded of the salient features of the endocrine system. The purpose of this appendix is to provide a brief introduction to, and a refresher course in, the essential feature at the endocrine system. This in turn will facilitate reader's comprehension of the central and major issues involved in contemporary animal biotechnology.

2 An overview of the endocrine system

The endocrine system provides a means of co-ordinating the metabolism of the various organs of the body. Essentially, therefore, the endocrine system is concerned with control. It is through the endocrine system that an animal can balance the needs of one part of its body, with the needs of another. For example when energy is rapidly required in muscle, fats and oil might be broken down in other parts of the body and transported from an area of 'energy surplus' to an area of 'energy need'.

hormones

The co-ordination between metabolism in one part of the body with that of another is achieved through the activities of a group of specific compounds called hormones. For the most part, hormones are produced in groups of cells called endocrine cells and are secreted into the neighbouring blood stream. A key feature of the endocrine cells is that they are sensitive to changes in their environment. The stimulus for an endocrine cell to secrete a hormone into its surroundings is however quite variable. Some are stimulated directly by chemicals (eg food), others are sensitive to an electrical pulse such as those produced by a nervous impulse from the brain, and yet others are themselves influenced by hormones.

secretory glands

The endocrine cells are usually grouped together into secretory glands. These glands may be distinctive separate structures (eg pituitary gland, thyroid gland) or maybe part of an organ carrying out other functions (eg kidneys, pancreas or intestinal tract). Figure 1 shows the main hormone producing sites. Although a human body is used to demonstrate the major organisms, the basic organ layout of domestic animals is entirely analogous. Note that the sexual organs have an endocrine function. This function is as important as their roles in reproduction. The hormones produced by the sexual organs can have a major impact on metabolic rates and growth, the levels of a wide variety of components (eg cholesterol) in blood and upon the morphological development of the rest of the animal. It will not be surprising to learn therefore that much of the biotechnology associated with animal husbandy is associated with the development of techniques to mimic or enhance the effects of these hormones.

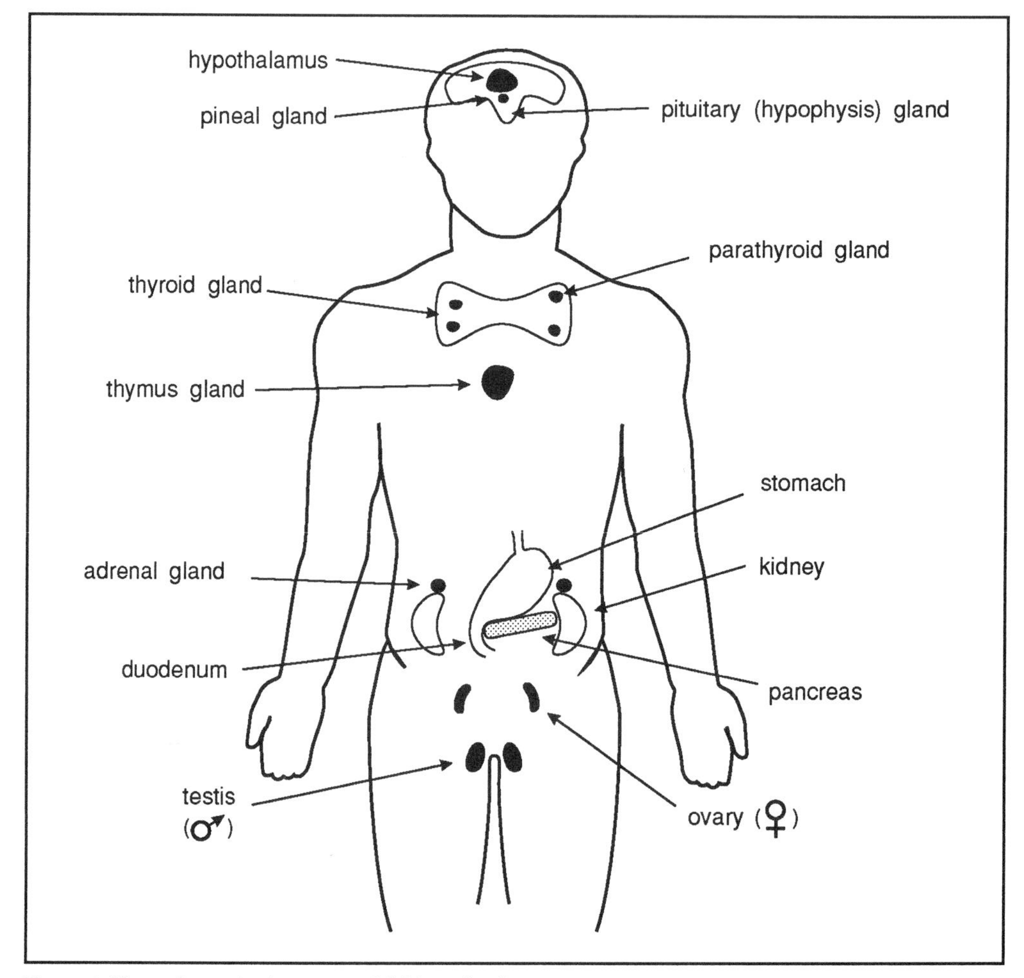

Figure 1 The major endocrine organs (highly stylised).

The implication behind Figure 1 is that the production of hormones is a function of specific groups of cells: There is also some specificity amongst the cells which respond to particular hormone action. This degree of specificity does however, vary. For example, muscle, liver and adipose (fat) cells all respond to adrenaline whilst thyroid cells are the only cells which are sensitive to thyroid stimulatory hormone (TSH). Note that some adipose cells cited above are also sensitive to insulin. The response of individual cells to a hormone may be dependent, therefore, not only on the amount of the hormone but also on the presence and concentration of other hormones. Whether or not a cell responds to a hormone depends upon whether it possesses the appropriate receptors. Receptors are protein-based structures on or in the target cells which have the capability of binding tightly to a particular hormone.

receptors

Hormone-secreting cells respond, to both environmental stimulii, and stimulii which arise within the body itself. Often internal and external stimulii have the opposite effect to each other. One stimulii switches on the secretion of a hormone, the other turns it off. A good example of this is the secretion of the adrenocorticotrophic hormone (ACTH) by the anterior lobe of the pituitary gland (Figure 2).

ACTH

Under the influence of an environmental stimulus such as traumatic stress, ACTH is secreted. ACTH secretion is turned off by cortisol produced by the adrenal cortex. This is however a rather simplified story. The environmental stimulus does not cause secretion of ACTH directly. The stimulus is transmitted by way of an ACTH releasing hormone (ACTH-RH) secreted by the hypothalamus. (This is often called corticotropin releasing factor, or CRF). You should also note that the secretion of cortisol occurs only in response to ACTH. Thus ACTH switches off its own secretion. Such a system is said to be a negative feedback system. When too much ACTH is released by the pituitary gland, this results in an increase in the release of cortisol from the adrenal glands which will thereby diminish the release of ACTH from the pituitary.

negative feedback system

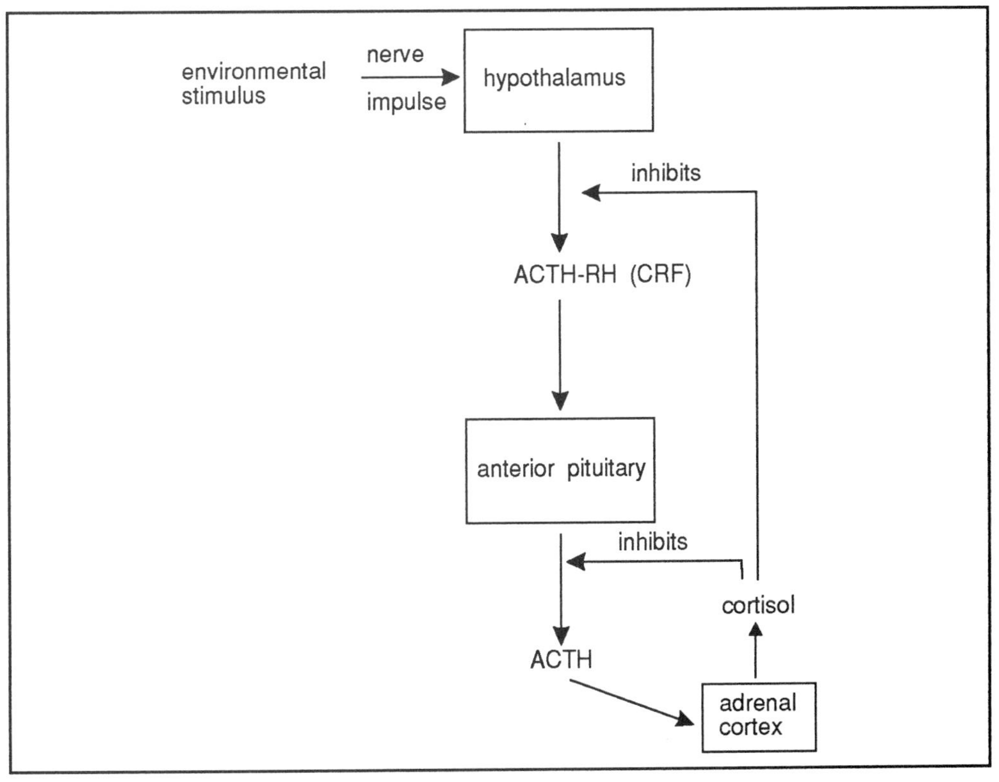

Figure 2 A negative feed-back control of hormone release (see text for further details).

There is one further feature of the endocrine system to be drawn from Figure 2. There are really two levels of control. One level is a rather coarse control. In the case illustrated in Figure 2 this operates at the level of the hypothalamus. The activity of the hypothalamus, in terms of ACTH levels to be found in the blood, is regulated by the balance between the stimulus received from the environment and the level of cortisol. If this was the only control that operated then the level of ACTH in the blood stream would be rather coarsely controlled. Fine control is maintained at a second level.

Π Inspect Figure 2 and decide how the level of ACTH in the blood stream is finely 'tuned'.

You should have come to the conclusion that the level of ACTH found in the blood is finely tuned by the inhibitory effect of cortisol on the anterior pituitary gland.

Negative feedback is a common feature of endocrine control. Figure 3 illustrates several examples of hormones produced by the pituitary gland. The pituitary gland is a multilobed structure. We can conveniently think of it in few portions, the hypothalamus, the anterior lobe, the posterior lobe and the pars intermedia. We will examine its function more closely later. Notice that the hypothalamus produces a variety of releasing hormones which stimulate the release of hormones by the anterior lobe of the pituitary gland. The posterior lobe and pars intermedia also produce hormones.

We have referred to the influence of the environment on hormone release. You should note that environmental influences usually affect hormone release by the release of hormone or hormone-like (mediators) agents by secretory neurones. We can therefore visualise a sequential chain of events.

environmental factor → detected by neurones → neurosecretion of locally active mediator → stimulation/inhibition of hormone secretion by glands

It is through such sequences that factors like day length or photoperiod can influence hormone release. This is the basis of regulating the breeding season in many animals. It is also through such a mechanism that the body responds to 'fear'. Certain neurosecretory cells respond to the nerve cell input by secreting the hormone noradrenaline. It is noradrenaline that causes the physiological responses to fear (eg quickening of heartbeat etc).

Π Which of the following hormones shown in Figure 3 are subject to negative feedback: ACTH, LH, FSH, TSH, GH?

Figure 3 illustrates that all those listed, except GH, are subject to negative feedback control. What Figure 3 does not show is that many of the hormones produced by the peripheral glands also inhibit the production of releasing hormones in much the same way as cortisol inhibits the production of ACTH-RH (CRF) as we illustrated in Figure 2.

Π What are the main target organs of the hormones released by the anterior lobe of the pituitary?

You may have simply listed those drawn in Figure 3 (ie adrenal cortex, gonads, thyroid gland etc). The key feature is that the main targets are themselves hormone producers.

trophic hormones

mitogenic

Many of the hormones produced by the anterior lobe of the pituitary are described as 'trophic (meaning nourishing) hormones'. This is because they stimulate cell division (ie that is they are mitogenic) as well as stimulating hormone production. Mitogenic

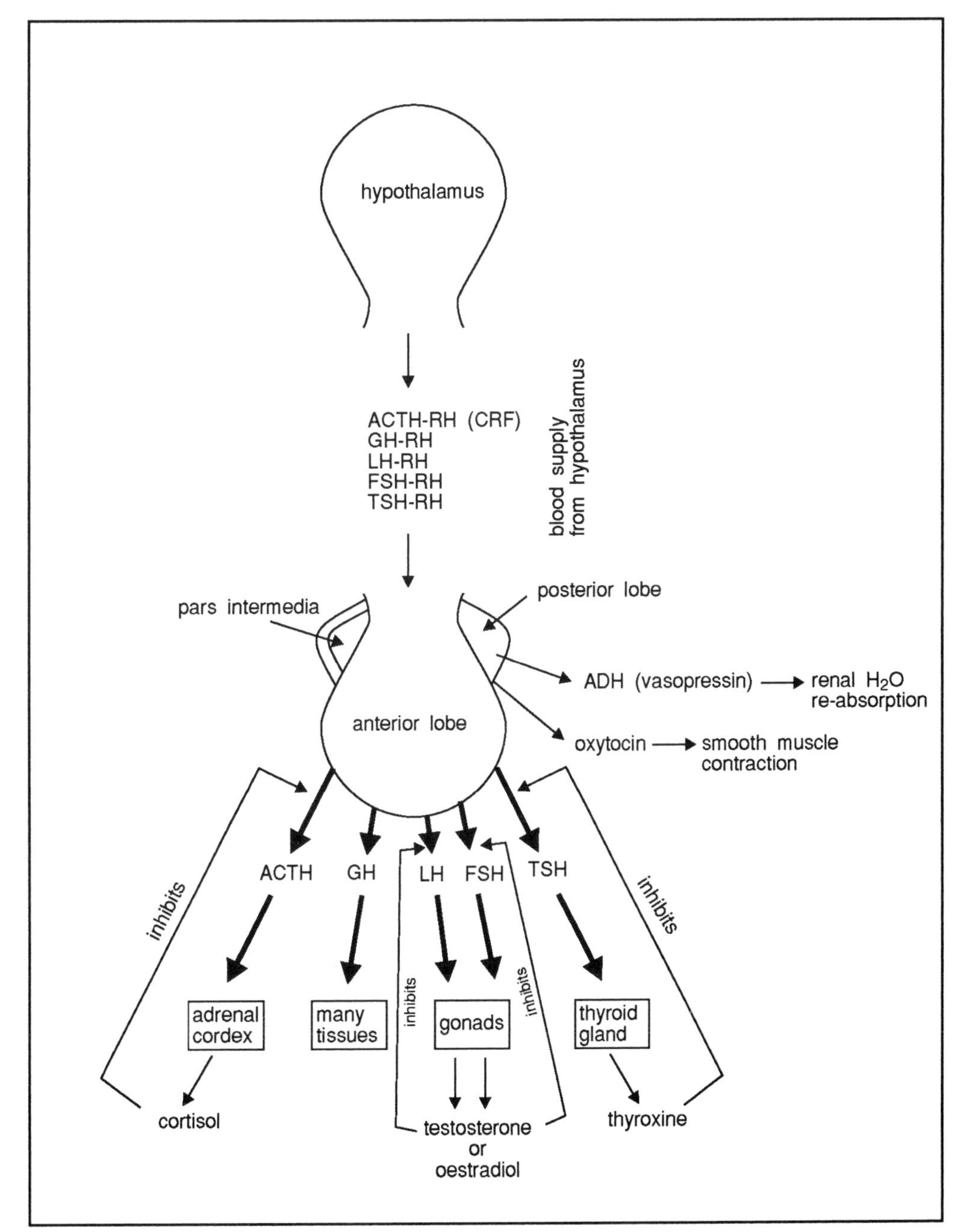

Figure 3 Hormone production by the pituitary gland. Key: ACTH = adrenocorticotrophic hormone, GH = growth hormone, LH = luteinising hormone, FSH = follicle stimulating hormone, TSH = thyroid stimulating hormone, GH-RH, ACTH-RH etc = releasing hormones, ADH = antidiuretic hormone.

(trophic) hormone production is not however confined to the pituitary. Some examples of trophic hormones are provided in Table 1.

Hormone	Target organ	Effect
Adrenocorticotrophic hormone (ACTH)	Adrenal cortex	Adrenal gland maturation secretion of corticosteroids
Thyroid stimulating hormone (TSH)	Thyroid	Secretion of thyroxine and other thyroid hormones
Luteinizing hormone (LH) Follicle stimulating hormone (FSH)	Gonads (ovaries or testes)	Maturation and secretion of oestragens and androgens
Growth hormone (GH)	Connective tissue (cartilage and bone)	Growth
Glucagon	β cells of endocrine pancreas	Secretion of insulin

Table 1 Example of trophic hormones.

SAQ 1

Indicate which of the following statements are true and which are false.

1) Growth hormone releasing hormone is produced in the pars intermedia region of the pituitary.

2) Smooth muscle contraction is stimulated by oxytoin.

3) Testosterone is the active ingredient which inhibits the release of TSH.

4) The production of ACTH-RH (CRF) is inhibited by cortisol.

Note that the responses to the questions in this Appendix are at the end of the Responses to SAQs section.

The picture that should have emerged from this section is the animal body containing many organs (or glands) producing hormones. The production of these hormones is often subjected to negative feedback and a kind of cascade of signals take place. In this the hypothalamus produces hormone releasing-hormones which stimulate the anterior lobe of the pituitary to release hormones. These hormones, in turn, activate target organs. The effects on the target organs are that they are stimulated to mature and to begin releasing the hormones.

Before we examine each of the major organs of the endocrine system and the hormones they produce, it is worthwhile examining some underpinning features of hormone action. In the next section we will briefly examine the structure and synthesis of hormones before exploring the mechanisms of hormone action. Finally we will briefly review the endocrine organs and the hormones they produce. The intention is not to provide details of the molecular and biochemical consequences of hormone action but to provide a working knowledge of the sites of production and general physiological function of the major hormones.

3 The structure and synthesis of hormones

Hormones can be classified into three main groups according to their chemical structure:

- polypeptide or protein hormones;
- steroid hormones;
- tyrosine-based hormones.

3.1 Polypeptide and protein hormones

pre-prohormone

These hormones consist of chains of amino acids varying in length from small peptides (eg oxytocin) to large polypeptides (eg growth hormone). Their synthesis begins on the ribosomes of the endocrine cell as a large inactive protein molecule known as a 'pre-prohormone'. (Use Figure 4 to help you follow this description).

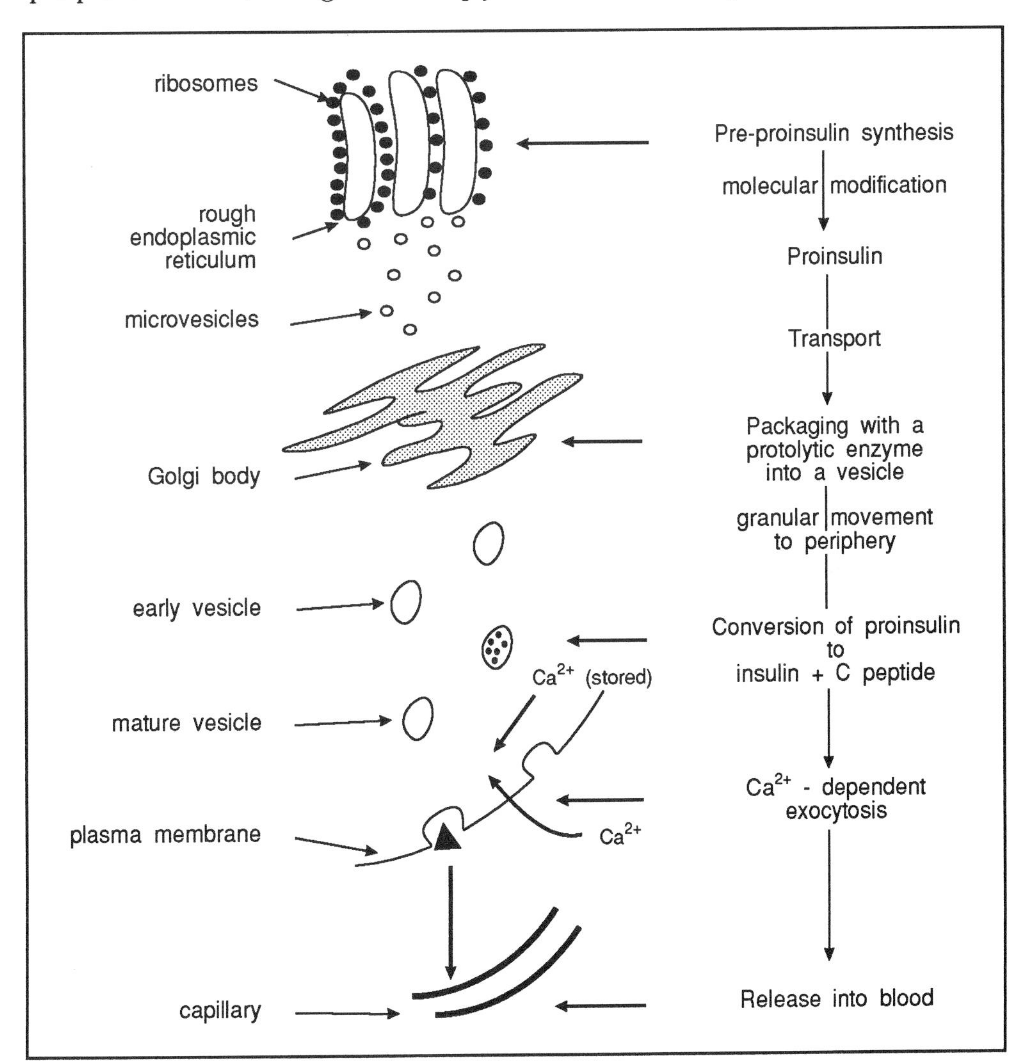

Figure 4 Polypeptide and protein hormone synthesis and release. The example shown is the production and release of insulin by the β cells of the pancreas.

prohormone

The molecular structure of this parent hormone is modified in the endoplasmic reticulum to form a smaller prohormone which is then packaged by the Golgi apparatus into a membrane bound vesicle containing a specific proteolytic enzyme. Further breakdown of the molecule occurs inside the vesicles to produce the smaller active hormone molecule and other peptide chains. This occurs whilst the vesicles migrate to the cell periphery where they are stored. Secretion of the hormone takes place by exocytosis, a process by which the contents of the vesicles are released into the extracellular fluid. This secretion is triggered by a rise in the concentration of calcium ions in the cytosol caused by opening of voltage sensitive calcium channels in the cell membrane or by release of calcium from intracellular stores. These two events, in turn, are brought about by the stimuli which excite that particular endocrine cell.

3.2 Steroid hormones

cholesterol desmolase

These hormones are all derived from cholesterol which is either delivered to the cell from the blood, manufactured metabolically or stored in the cell as lipid droplets. The cholesterol is first converted to pregnenolone by the enzyme cholesterol desmolase. This occurs in the mitochondria of the steroid producing cells (Figure 5).

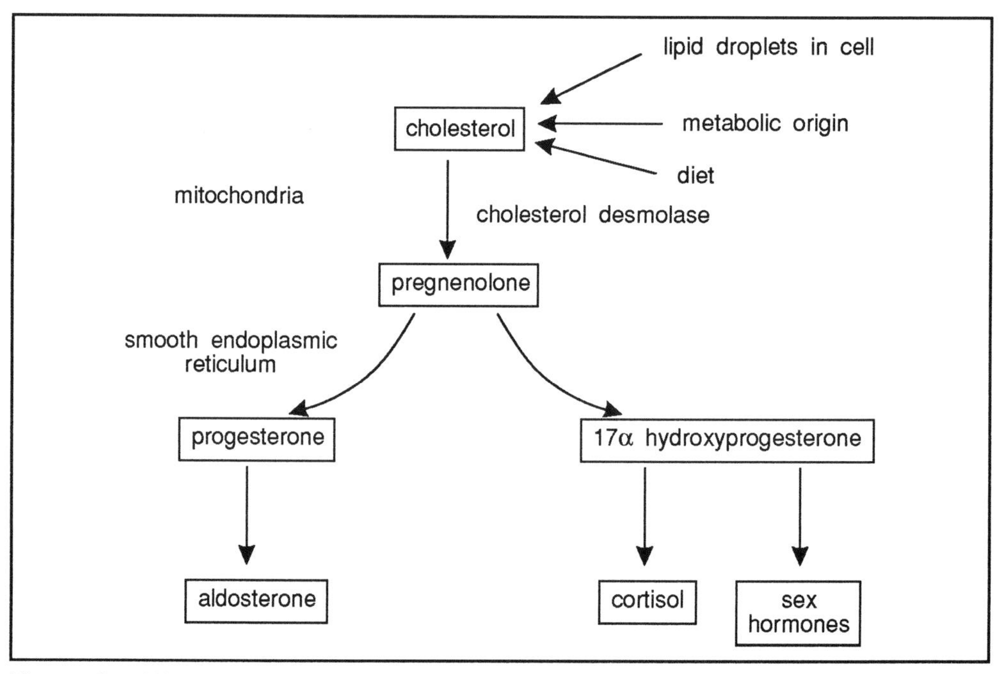

Figure 5 Steroid hormone production from cholesterol. Note that the final conversions occur in the mitochondria and smooth endoplasmic reticulum of the tissue involved. The final product depends on the specific enzyme systems present in the tissue.

fascicular and reticular zones of the adrenal cortex

The second stage takes place in the abundant smooth endoplasmic reticulum, characteristic of these cells, where the pregnenolone is converted to 17α hydroxyprogesterone or progesterone, the stages of synthesis beyond this point depend upon the presence or absence of specific enzymes which vary from tissue to tissue. The steroid hormone which is finally produced, therefore, depends upon the specific enzyme systems present in the endocrine cell eg cortisol is produced in the fascicular and reticular zones of the adrenal cortex which contain the enzyme 17α hydroxylase whereas aldosterone is manufactured in the glomerular zone which lacks this enzyme.

Unlike the peptide hormones, the steroid hormones are not stored in the cell. Once synthesised, they are secreted and, because of their lipid soluble structure, they diffuse across the cell membrane into the extracellular fluid and thence into the blood.

3.3 Tyrosine-based hormones

dopamine
adrenaline
noradrenaline
thyroxine
triiodothyronine

These hormones are derived from the amino acid tyrosine and include dopamine, adrenaline, noradrenaline (from the adrenal gland) and thyroxine and triiodothyronine from the thyroid gland. The synthesis of the catecholamines (noradrenaline and adrenaline) occurs by enzymatic modification of dopamine which is formed from the tyrosine molecule. This takes place in the chromaffin tissue of the body (the adrenal medulla) as well as in sympathetic adrenergic nerve endings (see Figure 6). The presence of a specific enzyme controls the final conversion of noradrenaline to adrenaline. As more of this enzyme is present in the adrenal medulla than in the adrenergic nerve endings it is found that the former tends to secrete four times more adrenaline than noradrenaline. In contrast, the nerve endings secrete mostly noradrenaline.

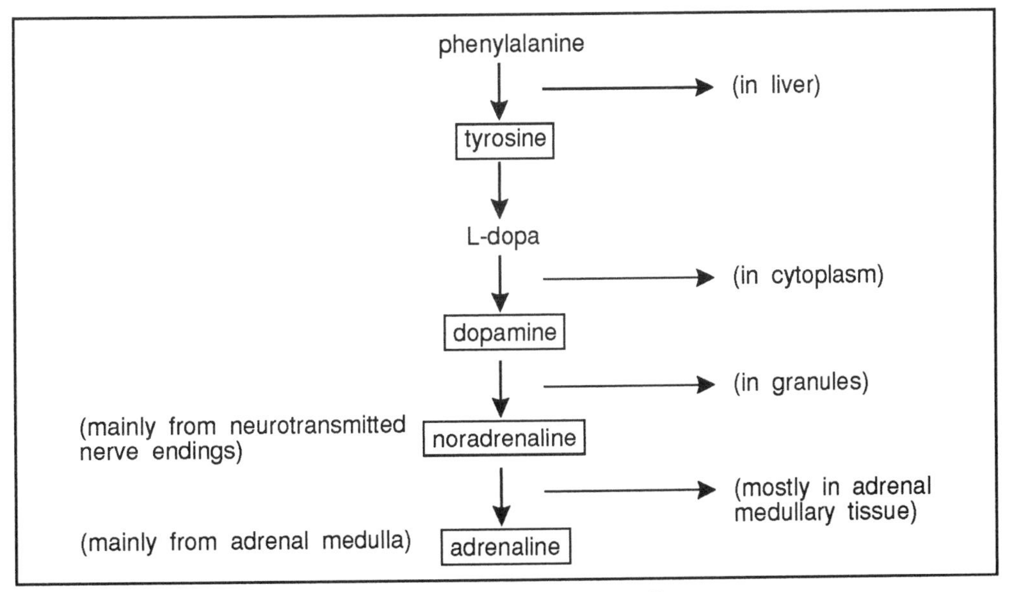

Figure 6 The production of adrenaline from tyrosine (see text for details).

chromagranin

The hormones are stored in granules with a carrier protein called chromagranin and their release is thought to be by exocytosis in a manner analogous to that involved in the secretion of peptide hormones.

The thyroid gland synthesises triiodothyronine (T3) and thyroxine (T4) by the iodination of tyrosine. This occurs in the thyroid follicles where the tyrosine residue of a protein, thyroglobulin (produced by the follicular cells) is iodinated by the addition of iodine in the presence of a peroxidase enzyme. The monoiodotyrosine and diiodotyrosine thus formed are condensed to form either of the two hormones. The release of the hormones is complex and results in the hormones being split from the thyroglobulin and transported from the colloidal centre of the follicles through the follicular cells into the blood. Figure 7 is a simplified diagram illustrating these events.

In Figure 7b, follow the path of iodine. Begin with the iodine in the blood stream and follow it through the cell to the colloid centre and then back, as triiodothyronine (T_3) or thyroxine (T_4) into the blood stream.

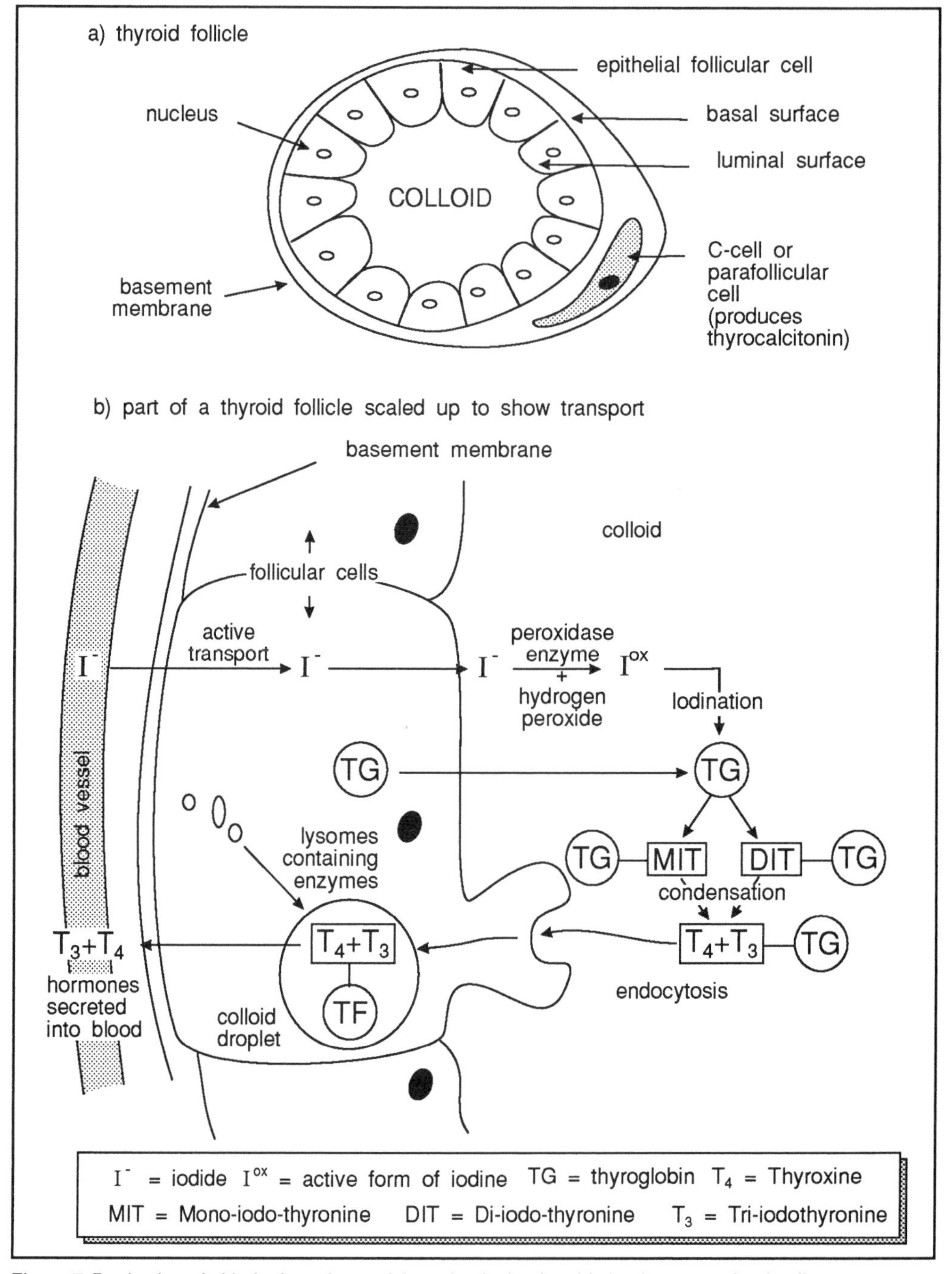

Figure 7 Production of tri-iodo-thyronine and thyroxine in the thyroid gland (see text for details).

Note that the colloid is a gelatinous substance containing the protein thyroglobulin (made and secreted by the follicular cells) and the thyroid hormones in all stages of synthesis. The oldest hormone is stored in the centre of the colloid; the most newly synthesised is at the periphery of the lumen (ie near the luminal surface of the follicular cells). The thyroid gland is composed of hundreds of these follicles interspersed with many blood capillaries.

SAQ 2

Put these phrases into the correct order to describe the synthesis and release of peptide hormones.

1) Influx of calcium causes release of hormone by exocytosis.

2) The prohormone is packaged into vesicles by the Golgi apparatus.

3) The secretory vesicles migrate to the cell periphery.

4) The final breakdown of the hormone from its inactive to active form by a protease enzyme occurs in the vesicles.

5) The pre-prohormone is formed on the ribosomes.

6) Modification of the molecular structure of the parent molecule takes place in the rough endoplasmic reticulum.

SAQ 3

Indicate which of the following statements are false. Give reasons for your answers.

1) Steroid hormones are all derived from cholesterol.

2) Steroid hormones are stored in the cells in which they are synthesised.

3) Peptide hormones are synthesised from large, inactive parent molecules called pre-prohormones.

4) Calcium ions are essential for the release of all types of hormones.

5) The molecular structure of thyroid hormones and catecholamines are both based on tyrosine.

4 Transport of hormones in the blood

binding to plasma proteins and receptor proteins

As the peptide hormones and catecholamines are water soluble molecules they are carried, mostly, in the blood dissolved in the aqueous phase of plasma. The steroid and thyroid hormones, however, are bound to plasma proteins because of their relative water insolubility. The binding occurs either to specific carrier proteins such as cortisol binding globulin or thyroid binding globulin or, more generally, to the plasma albumins. Such binding seems to prevent both renal filtration of the hormones and their enzymatic degradation in addition to serving as a reservoir of hormone available for slow release. The latter point can be understood by reference to Figure 8.

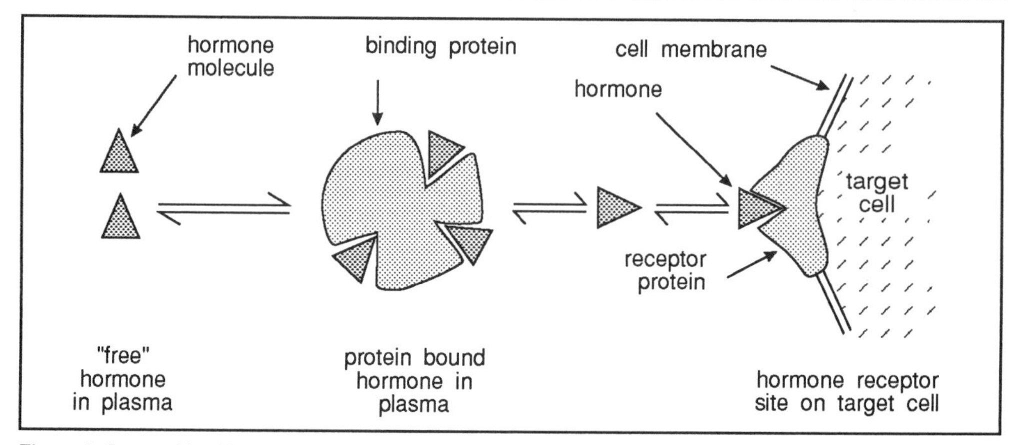

Figure 8 Competitive binding of hormones to plasma proteins and cell receptor proteins.

Here you can see that, although mostly bound to proteins, some hormone remains free (unbound) in the plasma and it is this fraction which is physiologically active ie brings about the response in the target cell. This unbound portion of hormones is always in equilibrium with the bound fraction. Once the free hormone interacts with the receptor on the large cell (which has a higher affinity for the hormone than does the plasma protein) then the equilibrium is disturbed and more of the bound hormone is released (dissociated) from the plasma proteins.

5 Metabolism and excretion of hormones

The amount of time a hormone remains in the bloodstream varies from a few minutes (eg antidiuretic hormone) to several days or weeks (eg thyroid hormones). This variability depends upon how the hormone is metabolised and excreted. The liver and kidney are the main routes for the removal of hormones but other sites such as the blood and the target tissues are also involved. Often the hormone is enzymatically degraded in the liver (bound hormones) or target tissue (free hormones) before being excreted in the urine.

An important point to note is that the metabolism of some hormones after their secretion is necessary to convert them into their active form. In other words, these hormones are secreted into the blood in an inactive form which is later metabolised to the active form, capable of bringing about the physiological responses of these hormones. This activation may occur in the blood (eg angiotensin I to angiotensin II) or in the target tissue (eg T4 to T3). Figure 9 summarises the various fates of hormones once they have been secreted.

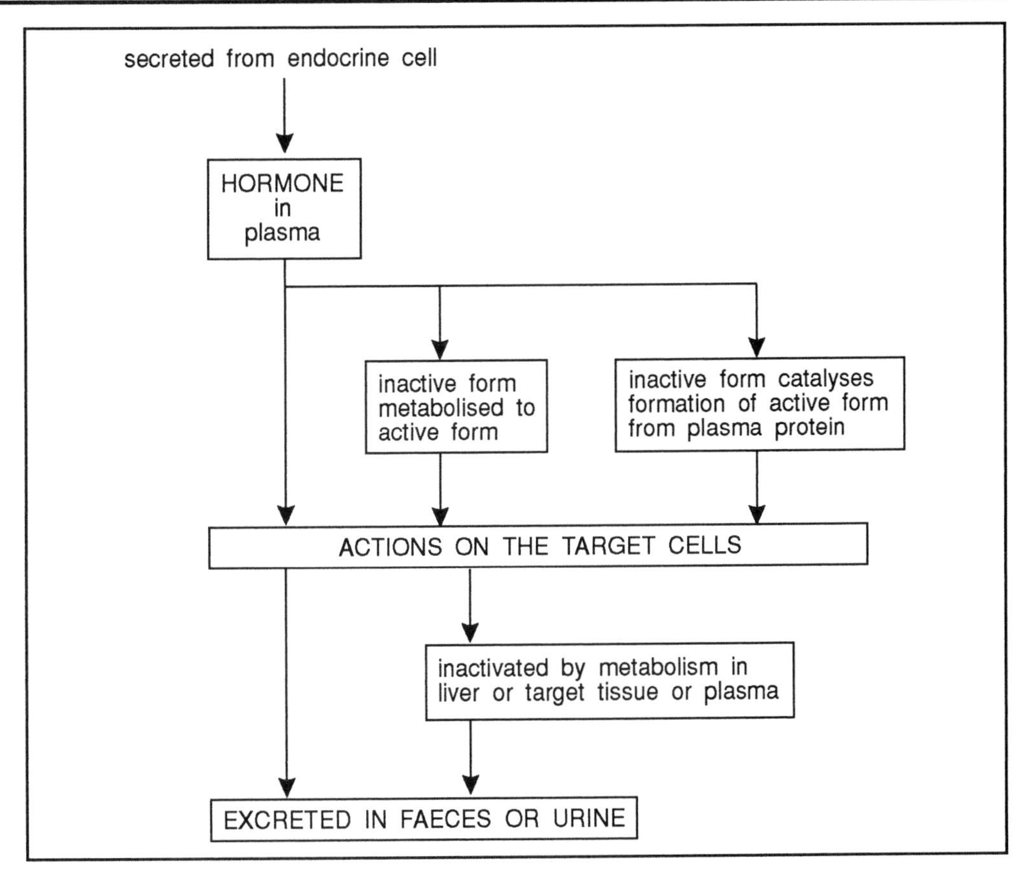

Figure 9 The inactivation and excretion of hormones.

Π Use Figure 9 to enable you to list three factors which determine the concentration of a hormone in the blood at any one time.

Your list should have included the following:

- the rate of secretion;
- the rate at which the hormone is broken down;
- the rate of excretion of the hormone.

These will all control the concentration of the hormone in the blood at any given time.

6 The mode of action of hormones

6.1 Receptors

Hormones are carried to virtually all tissues in the body by way of the bloodstream. However, not all of those tissues respond to all of the hormones. The response of the tissues is highly specific and involves only the target cells responding to a particular hormone. This ability to respond depends on the presence on, or in, the target cells of 'receptor molecules' which are specific for a given hormone. Thus the specificity of the target cell depends upon the presence of these receptors. That is to say, if a cell has a receptor for a particular hormone it will respond to that hormone. If it does not possess a receptor, then, although the hormone may be transported to a cell, the cell will not respond to that hormone.

receptor molecules

Π Why do you think thyroid gland cells will respond to thyroid stimulating hormone whereas ovarian cells will not?

You should have been able to suggest that ovarian cells do not contain receptor molecules able to bind thyroid stimulating hormone. They are, therefore, unable to detect and respond to this hormone. This is represented diagrammatically in Figure 10.

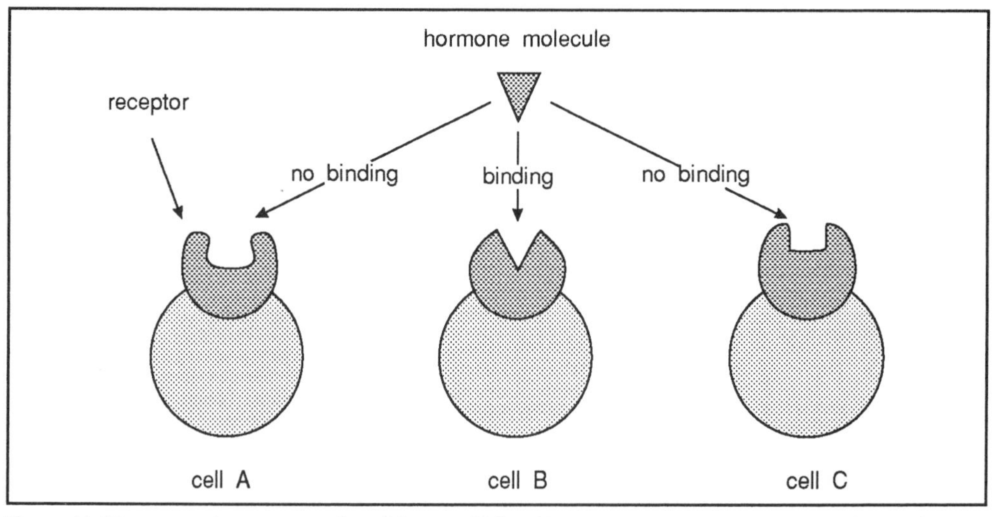

Figure 10 Diagrammatic representation of the specificity of hormone action. In the example shown, the hormone will only stimulate cell type B. Cell types A and C do not have receptors capable of binding the hormone' and therefore do not respond to its presence.

The receptor molecules are proteins which exist either in the cell membrane or in the cytosol or in the nucleus of the target cell. They function on a 'lock and key' basis in that the molecular shape of the receptors exactly matches the shape of the specific hormone molecule. Once binding occurs, the receptor-hormone complex acts as a 'switch' to initiate events inside the cell leading to the cell's response.

A cell may have more than one receptor type which may respond to either the same hormone or to different hormones. This means that a cell can give different responses to the same hormone (where the different receptors are acting as switches for different cellular events) or the cell may respond to more than one hormone. An example of the former behaviour is illustrated in Figure 14 where you can see that adrenaline produces two different responses from smooth muscle mediated either by an α-receptor or via a β-receptor.

6.2 Regulation of receptors

The actual number of a particular receptor type that exists on a target cell and the affinity of those receptors for their specific hormone are under physiological regulation. Where a cell is exposed for a prolonged time to a high concentration of a hormone, then the number of receptors on the cell for that hormone is decreased. This is known as 'down-regulation'. It has the effect of decreasing the cell's responsiveness to the hormone over a period of time, therefore, in effect, acting as a local negative feedback control.

down-regulation

Conversely, changes in the opposite direction, known as 'up-regulation', occur when a cell develops more receptors to a particular hormone in response to a prolonged exposure to low concentrations of that hormone. The result of this change is to increase the sensitivity of the tissue to that specific hormone. Physiologically, down-regulation of receptors is found to occur much more frequently than up-regulation.

up-regulation

This increase or decrease in receptor numbers is possible because there is a continuous turnover of the receptor molecules within the cells as a result of both degradation and synthesis of the constituent proteins. The changes, not only may occur as a result of physiological control, but also from pathological processes.

SAQ 4

Which of the following statements is the correct cause of down-regulation of hormone receptors?

1) When a receptor is exposed for a short time to high plasma levels of a specific hormone.

2) When a receptor is exposed for a prolonged time to low plasma levels of its specific hormone.

3) When a receptor is exposed to high plasma levels of its specific hormone for a prolonged length of time.

4) When a receptor is exposed to high levels of any hormone in the plasma.

6.3 Receptor Interaction

A final property of receptor molecules to be mentioned is the one which underlies the phenomenon of 'permissiveness' whereby the presence of one hormone is essential for, or potentiates, the response of the cell to a second hormone. This characteristic depends upon the fact that a hormone is able, not only to influence the number of its own specific receptors on a cell, but also influence the number of receptors for other hormones. This change may be either an increase or a decrease in the number of receptors for the second hormone. The phenomenon is illustrated in Figure 11. When hormone A results in an increase in the number of receptors for hormone B then hormone A is said to be permissive to hormone B. Often only small amounts of the permissive hormone are required in order for the full effect of the hormone B to be achieved. The example in Figure 11 shows that small amounts of the permissive thyroid hormones are required for the full effect of adrenaline upon adipose (fatty) tissue.

permissiveness

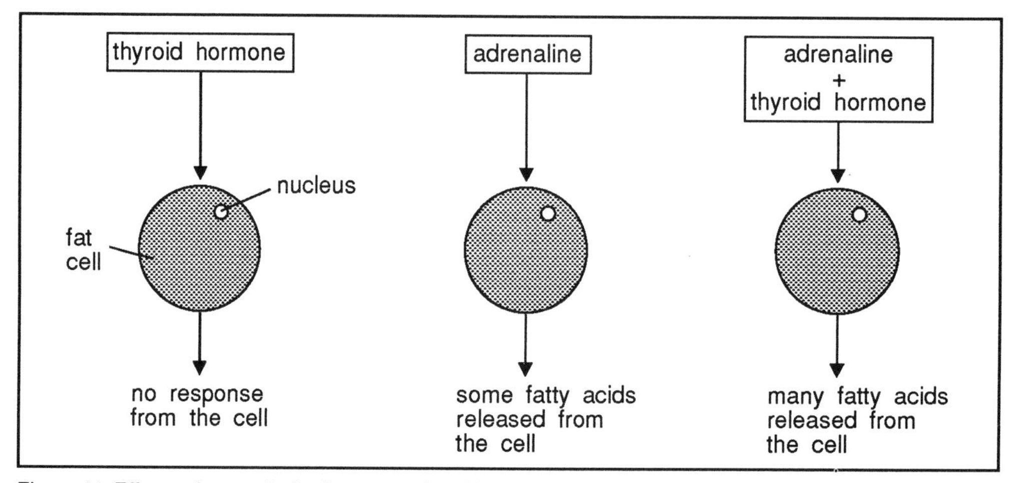

Figure 11 Effects of a permissive hormone (thyroid hormone) on the response of a fat cell to adrenaline.

Having considered the properties of the hormones themselves, we must now look at the way in which the intracellular events which constitute the cells' responses to hormones are triggered by the binding of specific hormones to their cellular receptors. These events are known as signal transduction mechanisms.

7 Events elicited by hormone-receptor binding: signal transduction mechanisms

7.1 Peptide hormones

second messengers

cAMP cGMP

inositol triphosphate diacylglycerol

For many hormones the intracellular events triggered by their binding to the plasma membrane receptors are mediated by intracellular molecules known as 'second messengers', the binding hormone being called the first messenger. Such systems are utilised by those hormones unable to penetrate the cell membrane because of their large size or their lipid insolubility. As one molecule of hormone can lead to the production of hundreds of molecules of second messenger, it is also important as a means of amplifying the original signal such that only minute amounts of hormones are required for physiological action. The most commonly known second messengers are cyclic AMP, cyclic GMP, calcium ion, inositol triphosphate and diacylglycerol. We shall consider these systems in turn.

Cyclic AMP

Cyclic 3′, 5′ -adenosine monophosphate (cyclic AMP, or simply cAMP) in the cytosol is derived from the high energy phosphate compound adenosine triphosphate (ATP). It is used as a second messenger by a great number and variety of hormones including, among others, thyroid stimulating hormone, adrenaline, antidiuretic hormone and luteinising hormone.

The binding of these hormones by their receptor molecules on the surface of the cell membrane results in an increase in the cytosolic concentration of the cAMP. How does this happen? The sequence of events is illustrated in Figure 12.

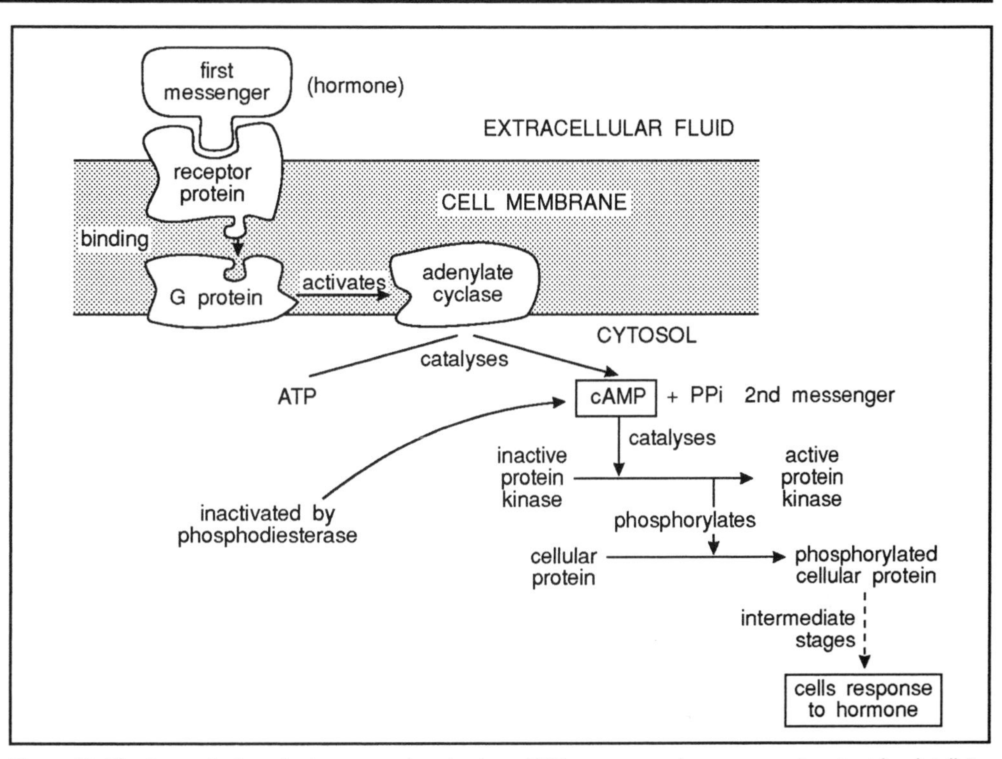

Figure 12 The transmission of a hormone signal using cAMP as a second messenger (see text for details).

G-protein

adenylate cyclase

The process begins when the binding of the receptor with the hormone causes a conformational change (a change in the molecular structure) in the receptor molecule which enable it to combine with a membrane regulatory protein called a G-protein. This G-protein, can be either an excitatory Gs-protein or an inhibitory Gi-protein. In the case of Gs-protein, this activates the membrane bound enzyme adenylate cyclase which functions to catalyse the conversion of ATP to cAMP in the cytosol. Hence, the hormone binding to the receptor results in a rise in the cytosolic concentration of cAMP. The cAMP then behaves as a second messenger in triggering a series of intracellular events which bring about the cell's ultimate response to the hormone. These events are mediated by protein kinases, present in the cytosol, which depend upon cAMP to be activated (cyclic AMP dependent protein kinases or A-kinases). These kinases phosphorylate other proteins, often enzymes, within the cell. Such phosphorylations alter the activity of these proteins and thereby bring about the changes in the cell's behaviour eg secretion or contraction. The great number of different dependent protein kinases in the cytosol allows for a variety of responses to occur in the cells cAMP as a result of hormone stimulation.

The effect of the cAMP is terminated by a cellular enzyme, called phosphodiesterase, which catalyses the breakdown of the second messenger to inactive non-cyclic AMP.

We should note, also, that the concentration of cAMP can be decreased, rather than increased, by some hormones. In these cases the hormones bind to receptors which combine with different regulatory proteins (Gi proteins) that inhibit, rather than activate, adenylate cyclase. This results in a fall in the concentration of cAMP and, therefore, fewer phophorylations of key proteins in the cell.

Cyclic GMP

Cyclic 3′, 5′ -guanosine monophosphate is another cyclic nucleotide which serves as a second messenger in a system analogous to cAMP. Cyclic GMP activates 'cGMP' dependent protein kinases which phosphorylate different cellular proteins from those activated by cAMP-dependent kinases and bring about alternative cellular responses. This system, however, does not appear to be as widely used as the cAMP second messenger system.

Π A hormone uses cAMP as its signal transduction mechanism but can elicit 5 different responses from its target cells. A drug is able to block just 1 of these responses. Explain whether the drug is blocking the action of the receptor, the G-protein, adenylate cyclase, cAMP or none of these stages in the second messenger pathway.

You should have come to the conclusion that, as all of these stages are common to all the responses produced by the hormone, then the drug cannot be blocking any of the above.

SAQ 5

Put the following phrases into the correct order to explain the cyclic AMP second messenger system.

1) The cAMP is inactivated by phosphodiesterase

2) The first messenger binds to the specific membrance receptor.

3) The activated protein kinase phosphorylates proteins involved with the cell's response to the hormone.

4) The regulatory protein (Gs) activates the membrane bound enzyme adenylate cyclase.

5) Adenylate cyclase catalyses the conversion of ATP to cAMP, resulting in an increase in the cytosolic concentration of cAMP.

6) The receptor/hormone complex binds to a membrane regulatory protein called a Gs-protein.

7) Cyclic AMP activates specific protein kinases in the cytosol.

Calcium ions and inositol triphosphate

Calcium ions are very important second messengers which functions in a great number and variety of cellular responses to hormones and other chemical messengers. As with cAMP, the key to their activity as a second messenger is a rise in the intracellular cytosolic concentration of the ion as a result of the hormone binding to its receptor at the cell membrane. This rise in cytosolic concentration of calcium ions may occur in several ways. Firstly, the binding of the hormone to the receptor may open up ligand-, or receptor operated calcium channels in the cell membrane allowing calcium to diffuse into the cell down its concentration gradient.

calcium channels

The calcium channels open up either because the receptor protein forms part of the channel and its conformational change on binding causes the channel to open, or because the receptor protein links up with a G-protein which, in turn, opens the channel up (see Figure 13). These different routes are labelled A and B in the figure. It should be noted, also, that these channels can be opened or closed by electrical events, rather than by chemical events, at the cell surface, in which case they are described as voltage gated channels.

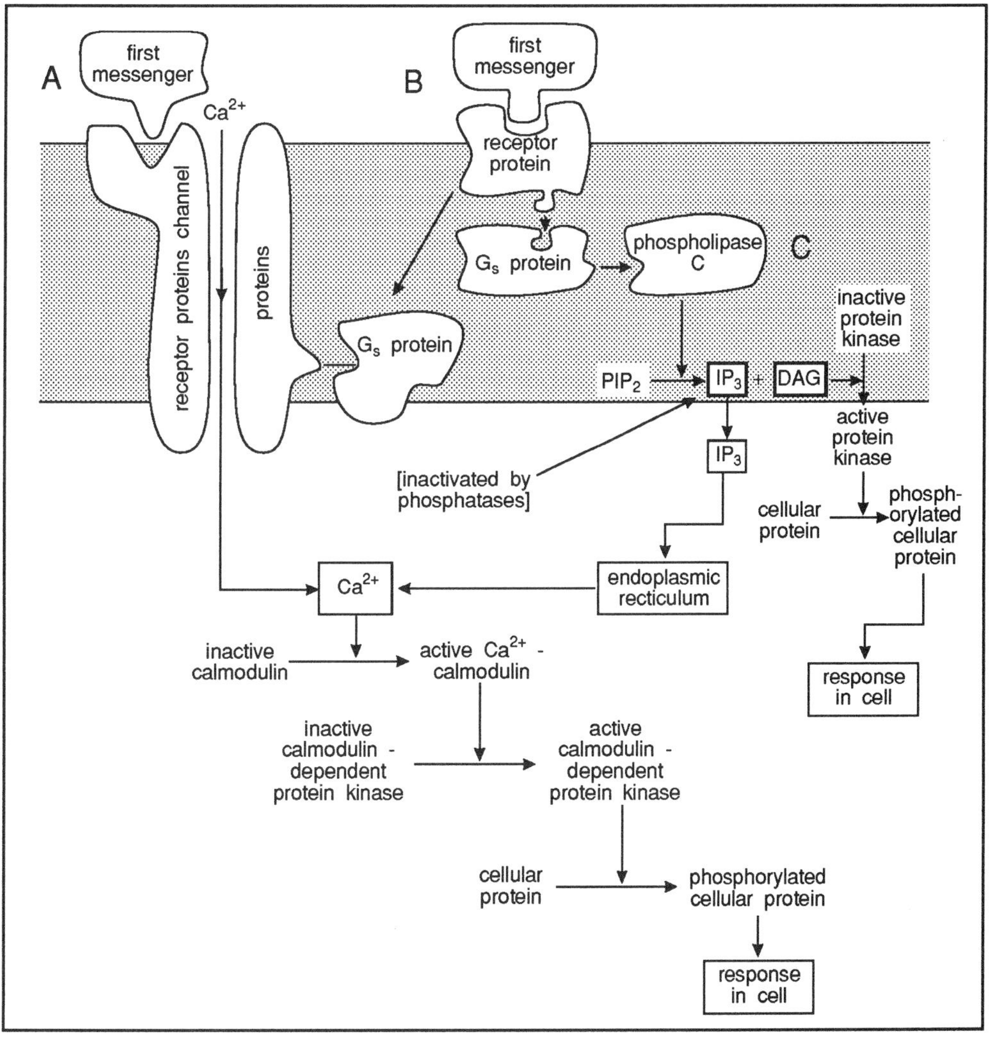

Figure 13 Ca^{++} and inosine triphosphate mediated regulation (see text for details).

Secondly, the cytosolic concentration of calcium ions can be increased by the release of calcium from intracellular stores. This occurs as a result of a different sequence of events which involves one of the phospholipid constituents of the cell membrane. In this sequence, the bound receptor activates a membrane bound enzyme called 'phospholipase C'. (Route C in Figure 13). The coupling of the receptor/hormone complex to this enzyme is mediated by another G-protein different from that involved in the cAMP pathway. The action of the phospholipase C is to catalyse the breakdown of a membrane phospholipid, phosphatidylinositol 4, 5-bisphosphate (PIP_2) to inositol

phopholipase C

triphosphate (IP_3) and diacylglycerol (DAG). The IP_3 enters the cytosol where it causes the movement of calcium ions from intracellular stores. The mechanism of this movement is, as yet, unknown, but evidence suggests the involvement of regulatory proteins within the membranes of the intracellular structures.

The effect of IP_3 is terminated by the action of specific cytosolic phosphatases which removes the phosphate groups leaving free inositol to be incorporated into the cell membrane. These events are illustrated in Figure 13.

Thus, these two mechanisms exist to increase the concentration of calcium ions in the cytosol ie opening up calcium membrane channels to allow entry of calcium ions from the extracellular fluid or the release of calcium ions from intracellular stores. However, it has been found that they often work in conjunction with each other rather than functioning exclusively. The two mechanisms may even be initiated by the same hormone.

SAQ 6

A hormone's response in a cell is mediated via a rise in the concentration of intracellular calcium ions. If a drug which blocked all the membrane receptor-operated channels for calcium were given, would it abolish the response of the cell completely? Explain your answer.

So, how does the raised intracellular calcium level lead to the eventual response of the cell to a hormone? The first step in this process is that the calcium binds, specifically, to calcium-binding proteins present in the cytosol. The most important of these is calmodulin. Calmodulin then activates or inhibits calmodulin dependent protein kinases which, in turn, lead to phosphorylation of proteins involved in the ultimate response of the cell to the hormone.

calmodulin

Other proteins similar to calmodulin function in the same way to mediate the second messenger actions of calcium. However, in some instances calcium ions may effect, directly, the cytosolic proteins involved in the cell's response without the mediation of specific calcium binding proteins.

Table 2 summarises the ways in which calcium acts as a second messenger.

The increase in calcium ion concentration in the cytosol can occur in two ways:

1) Opening up of plasma-membrane channels for calcium
2) Release of calcium from intracellular stores mediated by inositol triphosphate.

The calcium ions then act as second messengers in one of three ways:

1) They bind to calmodulin, which then activates a calcium-calmodulin dependent protein kinase whose function is to phosphorylate cellular proteins and, thereby, bring about the cellular responses.
2) Calcium binds with calcium-binding proteins other than calmodulin. The ensuing events are analogous to those above.
3) Calcium binds to the cellular (response) proteins directly without any mediatiors.

Table 2 Ways in which calcium ions act as a second messenger.

A further consideration of the involvement of inositol triphosphate (IP_3) in this system is that recent evidence has shown that the IP_3 can be phosphorylated further to form inositol tetraphosphate (IP_4). This molecule appears to act as an intracellular messenger to open membrane calcium channels which allow calcium ions to enter the cell and, either replenish the stores depleted by the IP_3, or, perhaps, be redistributed within the cell.

Diacylglycerol (DAG)

PIP_2

The final second messenger to be considered is the molecule 'diacylglycerol' produced from the breakdown of PIP_2 by phospholipase C (see last section). The DAG, as a lipid, remains in the cell membrane where it activates a specific protein kinase, called protein kinase C, which is capable of phosphorylating a large number of other proteins in the cell. It appears to be very important in the down-regulation of membrane protein receptors, as it phosporylates these receptors thus rendering them ineffective. (Cyclic AMP dependent protein kinases seem to function in a similar manner).

prostaglandins

The DAG is inactivated either by specific protein kinases and reincorporated into the cell membrane or by a lipase, in which case, arachidonic acid is liberated. This latter molecule is the precursor of the prostaglandins, a further group of local messenger molecules.

It is important to emphasise that a hormone initiating events via IP_3 will, simultaneously, release DAG and, therefore, will activate protein kinase C. There is evidence, in many cases, that the molecules act synergistically ie they are both required in order to initiate the cell's response to a hormone. The use of various pharmacological agents have shown that if one or other of the molecules is inactivated, then the cell's response to the hormone may be blocked.

Not only is there a functional link between IP_3 and DAG but also there exist important interactions between all the different second messengers. They frequently function together, sometimes causing effects which are in opposition to each other, other times the changes are in the same direction. Figure 14 shows how effects of adrenaline on smooth muscle are mediated by two different signal transduction mechanisms.

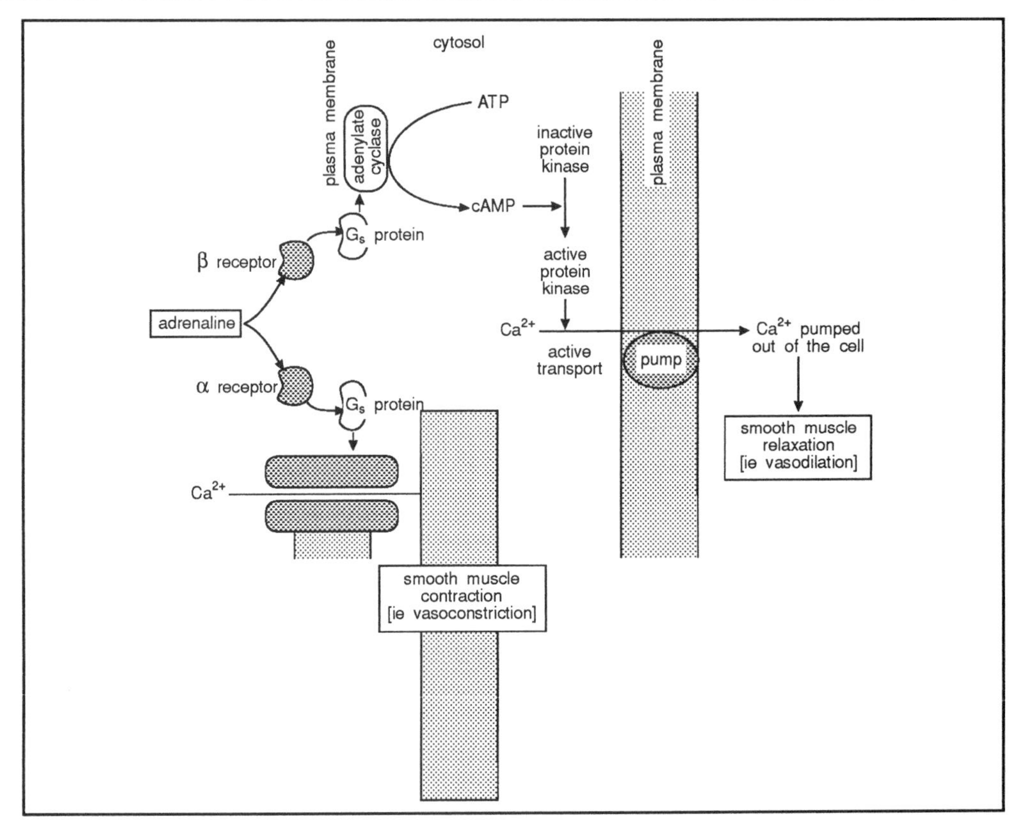

Figure 14 Interaction between second messengers. The interaction is illustrated using adrenaline which can lead to increases in the cellular levels of cAMP and Ca^{2+}. Note that cAMP can also influence the cellular levels of Ca^{2+}. The final response of the blood vessel to adrenaline depends on the comparative numbers of α and β receptors.

Receptors acting as protein kinases

So far we have described four molecules which can activate protein kinases - cAMP, cGMP, calcium-activated calmodulin and DAG. To finish this section we must consider one further mechanism by which the first messenger can trigger the activation of protein kinases. This mechanism involves the receptor itself. When the first messenger binds to the receptor, the receptor then becomes the active protein kinase which phosphorylates either itself or other cytosolic or membrane proteins. These protein kinases, as a group, are referred to as tyrosine kinases because the site of phosphorylation is the tyrosine portion of the target protein. An example of a hormone which is thought to function in this manner is insulin.

tyrosine kinases

A final comment to be made in this section is that a particular receptor type is often associated with a particular signal transduction mechanism in a variety of cells. The different events triggered by this mechanism result in the various responses of the cells.

SAQ 7

Which of the following statements is incorrect. Give reasons for your answers.

1) IP_3 acts to increase the cytosolic concentration of calcium ions by opening up membrane channels for calcium.

2) DAG activates protein kinase C.

3) Phosphodiesterase inactivates both cAMP and DAG.

4) The same regulatory G-protein mediates transduction of all signals across cell membranes.

5) A hormone initiating the activation of the IP_3 pathway will simultaneously cause an increase in the concentration of DAG in the cell membrane.

SAQ 8

You are presented with the problem of identifying the second messenger systems involved in three different responses (A, B, C) of a particular cell type to a hormone. You are given 3 pharmacological agents to manipulate the pathways:

1) Forskolin which activates adenylate cyclase.

2) A calcium ionophore which causes an increase in the concentration of calcium ions in the cytosol.

3) Theophylline which inhibits phosphodiesterase.

4) Imidazole which activates phosphodiesterase.

Your result are summarised in the table below:

	A	B	C
forskolin	100	0	25
ionophore	0	100	25
theophylline	100	0	25
imidazole	0	0	0
theophylline + ionophore	100	100	100

The table shows in arbitrary units the results of pharmacological manipulation of cellular responses A,B and C.

From these data try to surmise which second messenger system is the mediator in each response.

7.2 Steroid and Thyroid Hormones

Steroid and thyroid hormones share a different signal transduction mechanism. These hormone molecules are able to penetrate the cell membrane, steroids because of their lipid solubility and thyroid hormones because they are transported into the cell.

cytosolic receptors

The specific receptors for these hormones are found in the cytosol or in the nucleus. In the case of the cytosolic receptors, the hormone molecules bind to the receptor and the complex is translocated into the nucleus of the cell (see Figure 15). Once inside the nucleus the receptor-hormone complex binds to the chromatin, stimulating the production of specific mRNA. The mRNA then enters the cytosol to act as a template for the synthesis of cellular proteins involved in the cell's response to the hormone. These events follow a longer time course than the second messenger systems such that steroid and thyroid hormones are characterised by a typical lag period of 45 mins. before a response in the cell is apparent. This contrasts with the virtually instantaneous effects of the other second messenger systems.

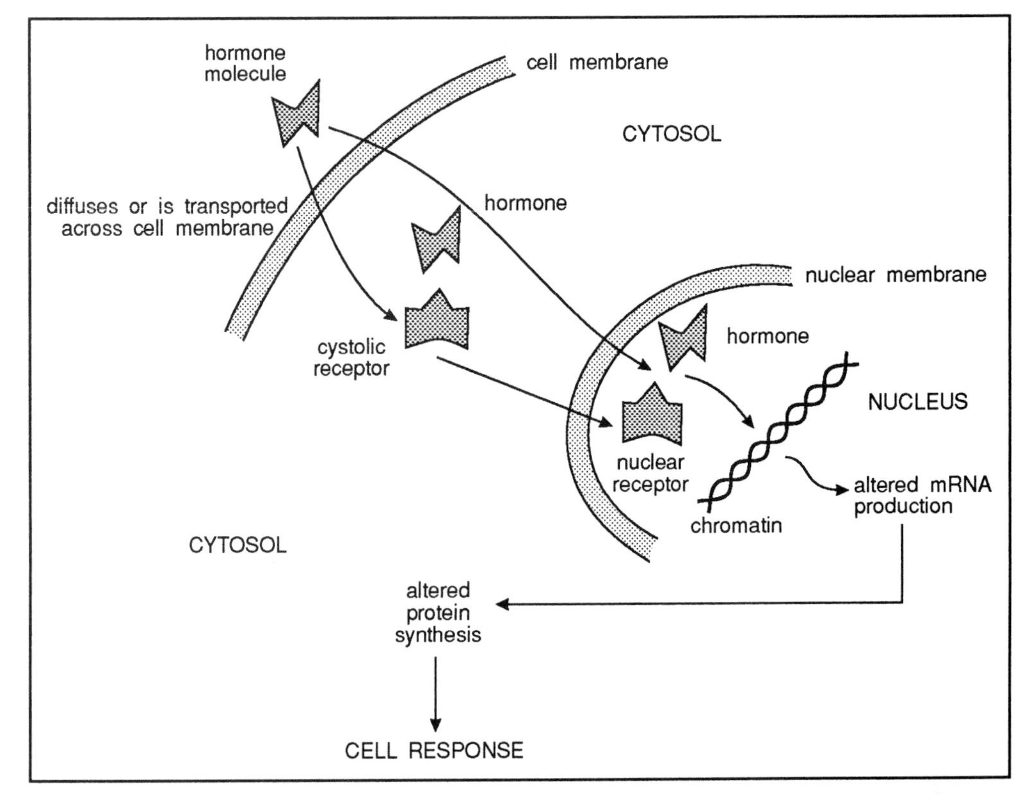

Figure 15 Diagramatic representation of cytoslic hormone receptors and their migration into the nucleus (see text for further details).

Π Actinomycin-D is an antibiotic which blocks RNA synthesis. What affect would this have on the action of a steroid hormone?

You should have concluded that this antibiotic will probably block the action of steroid hormones. This is because the steroid-receptor complex works by altering the nature of gene expression (ie transcription). Since this involves RNA synthesis, actinomycin D would block the action of the steroid-receptor complex.

8 Control of hormone secretions

8.1 Rhythm

circadian rhythm

Many hormones are not secreted continuously. Instead, their fluctuating plasma concentrations follow a variety of rhythmical patterns. Amongst others, these include the occurrence of short bursts of output every few hours (growth hormone), a 24-hour cyclical variation (circadian rhythm eg cortisol), a monthly rhythm (gonadal hormones) or seasonal variations in output (pineal hormones).

Often, such rhythms result from the influence of external factors such as sleep, light/dark cycles, environmental temperature and physical or mental shock, which are mediated by the nervous system.

The secretion of hormones is usually homeostatically controlled so that there is neither oversecretion nor undersecretion of the hormone as related to the needs of the body. When such regulation goes wrong, then pathological states ensue.

8.2 Feedback control

negative feedback

As we learnt earlier, hormone output is controlled by negative feedback. In such a feedback a change in a given parameter results in a response in the opposite direction in order to return that parameter to its original value. With some hormones there is a simple feedback control that involves the blood levels of an inorganic ion or organic substance which the hormone controls, eg insulin, which is controlled by the level of glucose in the blood or parathyroid hormone, which is controlled by the calcium levels in the blood. In the case of insulin: a rise in blood glucose stimulates the output of insulin from the pancreas, the insulin then acting to encourage the cellular uptake of the glucose which has the effect of lowering the levels in the blood. The decreased blood glucose levels then remove the original stimulus to the insulin output which, therefore, falls.

positive feedback

In a few instances, however, the hormone output is controlled by a positive feedback system whereby the controlling input increases the output even further, rather than reducing it as would a negative feedback control. This is the case for both the output of oxytocin in parturition and the output of luteinising hormone in the ovulatory surge. During parturition oxytocin is released as the foetus descends into the birth canal. The oxytocin causes uterine contractions which push the foetus further down, causing more output of oxytocin. This continues until the birth of the infant is complete.

In the early oestrous cycle, the rising levels of oestrogen exert a negative feedback on the output of luteinising hormone (LH) from the pituitary gland. However, once oestrogen levels reach a certain level in the plasma which is maintained for a given length of time, then this feedback suddenly becomes a positive feedback leading to a surge in the output of the LH and the release of the ovum from the ovary (ovulation). The mechanism for the sudden reversal of the feedback is, as yet, unknown.

8.3 Nervous control

The output of other hormones is controlled, not by blood parameters, but by the nervous system. This can be a direct effect of neurotransmitters released from nerve endings supplying the gland, as seen in the output of adrenaline from the adrenal gland which is stimulated by the excitation of sympathetic nerves. On the other hand, some hormones are released directly from nerve endings by the process of neurosecretion. The two well known examples of this are antidiuretic hormone (ADH) and oxytocin, both of which are neurosecreted from the posterior pituitary as a result of excitation of hypothalamic neurosecretory cells whose axons terminate there.

multiple inputs regulate hormone output

The other influence of the nervous system on endocrine output is via the hypothalamus and anterior pituitary gland which are important for the control of the majority of hormones in the body. Before considering this system in more detail, we should mention one or two other points on the control of hormone output. Firstly, that hormones are capable of influencing the output of each other and, secondly, that the output of a given hormone is often a result of more than one of these controlling inputs. In many instances it is a combination of blood parameters, nervous input and plasma levels of other hormones which influence the final output of a given hormone.

8.4 The Hypothalamus and Pituitary

The key to the function of the hypothalamo-pituitary influence on endocrine output is the process of neurosecretion. Neurosecretion is the manufacture and secretion of peptide hormones by specialised neurones, called neurosecretory cells. Such neurones exist in the hypothalamus, in two main areas, from which their axons influence the pituitary gland.

major areas of the hypothalamus and pituitary

The pituitary gland (hypophysis) is connected to the hypothalamus by the pituitary stalk (the infundibulum) and lies at the base of the brain. It is divided into two main sections, the anterior or glandular portion (the adenohypophysis) and the posterior or neural part (the neurohypophysis). In some animals there is a marked third section, the intermediate lobe, pars intermedia, but this is very small in humans.

The posterior pituitary is a downgrowth of the hypothalamus and is composed of the long axons and axon terminals of the neurosecretory cells situated in two distinct areas of the anterior hypothalamus. As described in the last section, the hormones ADH and oxytocin are manufactured in these cells, transported down the axons to be stored in vesicles in the axon terminals. The hormones are released by exocytosis as a result of excitation of these nerve cells by a variety of synaptic inputs. The process is summarised in Figure 16.

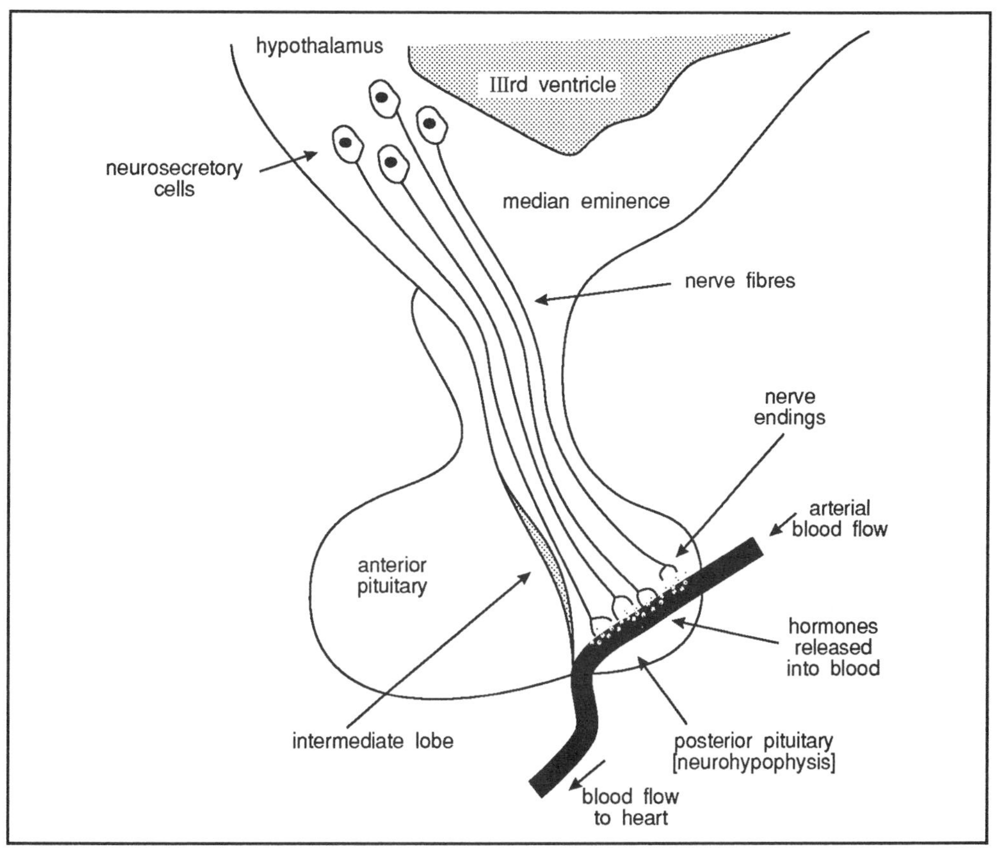

Figure 16 Release of hormones from the posterior pituitary (see text for details).

The anterior pituitary, however, is a glandular tissue developed as an outgrowth of the embryonic foregut. It is connected to the hypothalamus by a plexus of blood capillaries (the primary plexus) which drains the hypophysial portal vessels passing down through the infundibulum to the anterior pituitary, (see Figure 17a). Neurosecretory cells from another, less well differentiated, area of the hypothalamus send short axons to terminate in the region of the primary plexus at the top of the pituitary stalk. These cells produce a series of chemical messengers called hypothalamic releasing or regulatory hormones or factors. The latter distinction depends upon whether the chemical structure of the molecule is known (hormone) or not (factor). These factors may be either excitatory or inhibitory. They are released into the capillaries of the primary plexus from which they are transported via the 'portal vessels' to the glandular cells of the anterior pituitary, (see Figure 17b). This direct blood route to the adenohypophysis avoids the hormones being diluted or metabolised as they would be if they were secreted into the general circulation for distribution by the heart. When the hypothalamic releasing factors arrive in the anterior pituitary, they cause the gland cells to release their own stored hormones. There are six different anterior 'pituitary hormones', each synthesised by different types of cells. These hormones, in turn, are transported to their target endocrine glands where they stimulate the secretion of the target gland hormones. These hormones, then bring about a response from their specific target tissue.

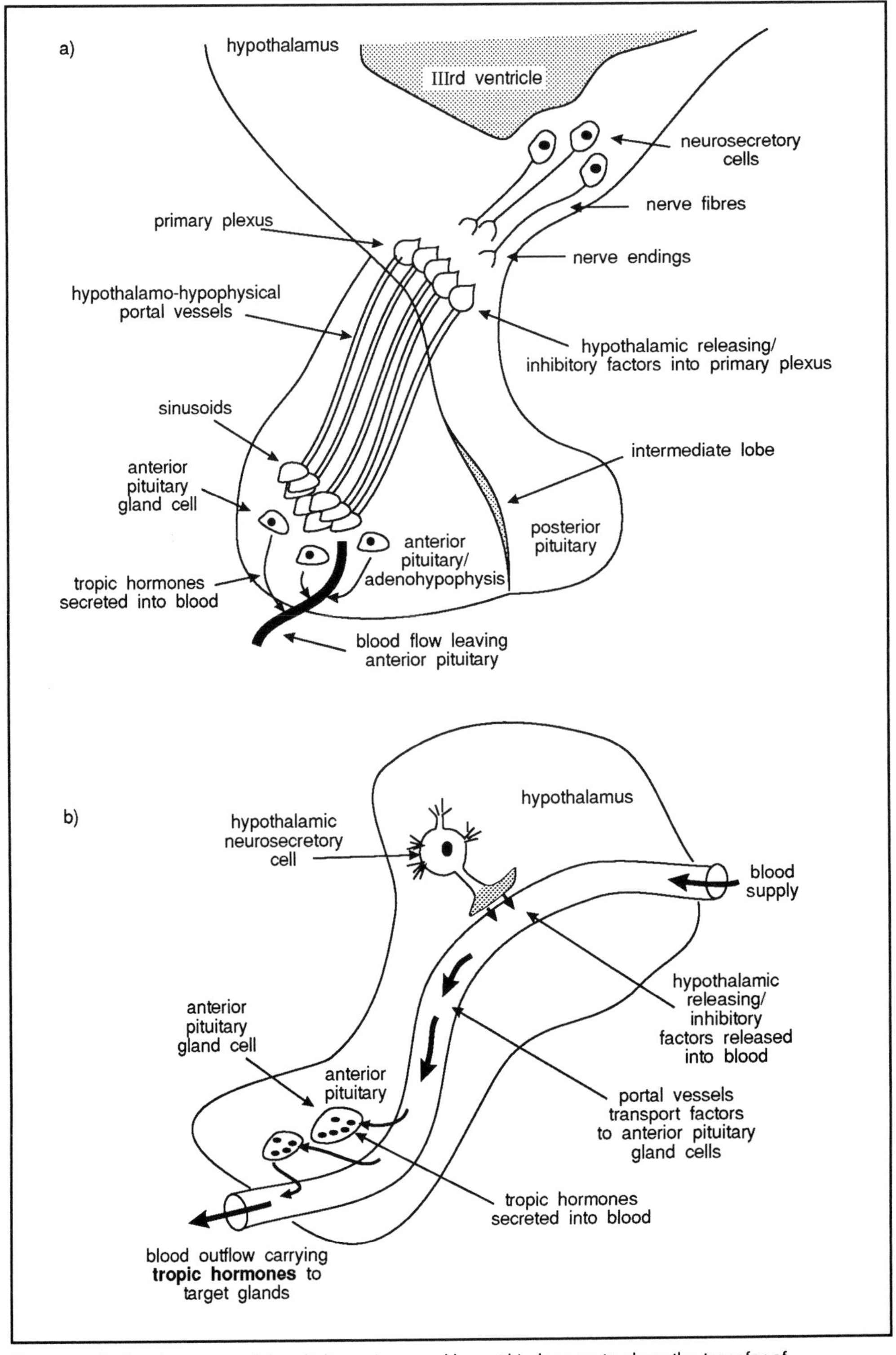

Figure 17 Stylised structure of the pituitary a) general layout b) close up to show the transfer of hypothalamic hormones to the anterior pituitary via the portal vessels.

The whole cascade of events is controlled by various negative feedback loops and from higher centre inputs to the hypothalamic neurosecretory cells. Figure 18 illustrates this control. The main feedback is from the level of the target hormone in the blood to the pituitary cells producing the tropic hormone. If the level of the target hormone becomes too high, the output of the tropic hormone will be reduced, resulting in less stimulation of the target endocrine gland and, therefore, a decrease in the output of the target hormone.

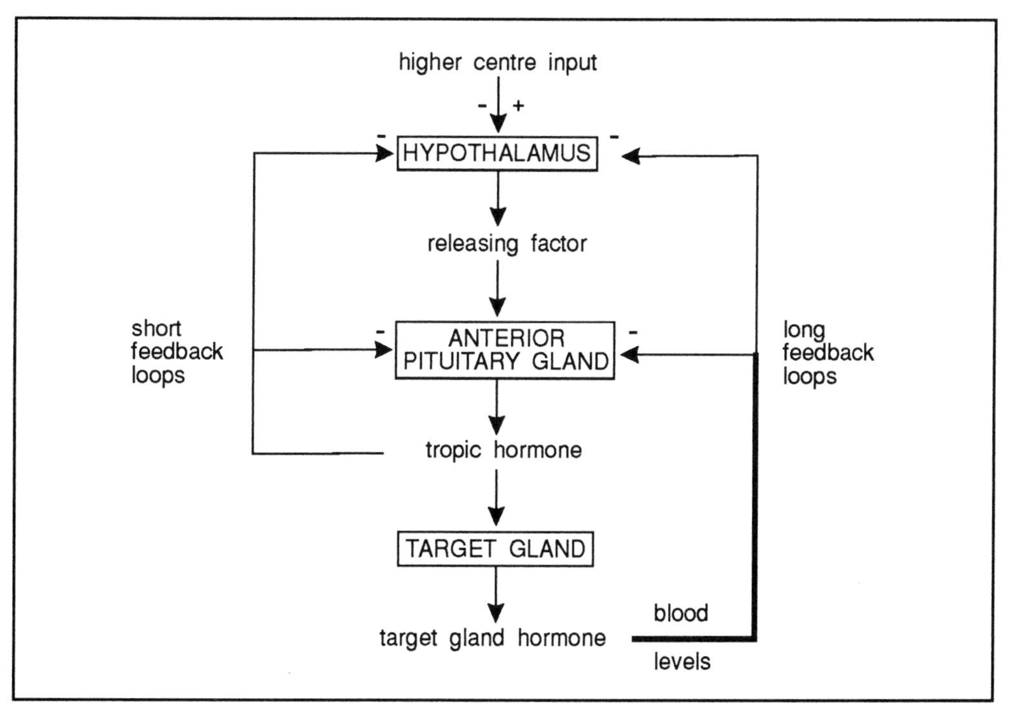

Figure 18 The sequence of events leading to the release of target gland hormones and their regulatory loops. Similar feedback loops exist for the hypothalamic inhibitory factors.

This feedback loop is thought to extend also to the hypothalamus to reduce the output of the releasing factor thereby, further decreasing the output of both the tropic and target hormones. (Such feedback control will also function for the inhibitory factors).

long-loop feedback

These feedback effects of the target hormone on the pituitary and hypothalamus are described as long-loop negative feedback controls.

short-loop feedback

Some of the pituitary hormones do not have a specific target endocrine gland whose output can regulate their secretion. For instance, prolactin which affects mammary tissue and growth hormone which affects all tissues in the body. The negative feedback in these cases, therefore, is via a short-loop feedback in which the pituitary hormone itself, exerts a negative feedback onto the output of the hypothalamic neurosecretory cells. The hormones, thus, regulate the output of their own releasing factors. Such short-loop feedbacks may also exist for the other pituitary hormones, thus adding a further refinement to the control of their output. In addition, recent evidence indicates that there may be some interaction between the various pituitary hormones and their feedback loops. It is thought that plasma levels of one hormone may influence the sensitivity of several pituitary cells to their hypothalamic releasing factors in addition to those regulating their own output.

SAQ 9

Draw a flow diagram to summarise the control of output of thyroid hormones from the hypothalamus and pituitary. The hypothalamic releasing factor is called thyroid releasing hormone and the pituitary tropic hormone is called thyroid stimulating hormone.

8.5 Other hormone secreting organs

Adrenals

medulla
cortex

The adrenals are two small organs situated above the kidneys. In humans, each are approximately 10 g in weight and consist of two parts, an inner medulla and an outer cortex.

The cortex secrets several different steroid hormones. These can be divided into two groups according to their function, the glucocorticoids (eg cortisol and the mineralo-corticords (eg aldosterone).

The medulla secretes adrenaline and noradrenaline (also referred to as epinephrine and norepinephrine).

The glucorticoids (cortisol) acts on many tissues and generally cause the degradation of protein and the utilisation of glycogenic amino acids to produce carbohydrates. Cortisol also has powerful anti-inflammatory action. The level of this hormone is regulated by ACTH (adrenocorticotrophic hormone), also called corticotrophin.

The mineralocorticords (eg aldosterone) regulate the K^+ and Na^+ content of body fluids. These hormones stimulate the re-absorption of Na^+ and the excretion of K^+ by the kidneys.

Adrenaline secretion is under control of the nervous system. Sympathetic nerve fibres controlled by the hypothalamus and other areas of the brain stimulate the release of these catecholamine hormones by the adrenal medulla. Adrenalin has a promoting effect on the degradation of carbohydrate and fat resulting in making glucose and fatty acids available to muscle. It also stimulates the heart rate and blood flow to the major organs.

Gonads

androgens

testosterone

We will consider the male gonads first. The testes secrete several steroids hormones (androgens) of which testosterone is most important. Testosterone maintains what are usually referred to as the secondary male characteristics (eg in human stature, deep voice, facial hair). The release of testosterone is under the control of the luteinising hormone (LH). The production of sperm by the testes is controlled by the follicle stimulating hormone (FSH).

oestrous cycle

The ovaries are also controlled by FSH and LH. The steroid hormones released by the ovaries (oestrogens) include oestradiol and oestrone. Oestrogen secretion is stimulated by LH. The function of the oestrogens is to control the oestrous cycle. They maintain the secondary female characteristics (eg in humans, stature, enlarged breasts, body fat).

corpus luteum

It is worthwhile here to briefly follow the events of the oestrous cycle. During this cycle, a follicle containing an ovum develops under the stimulation of FSH. This follicle produces oestrogens, which cause the walls of the uterus to thicken. For some time after the onset of menstruation, the secretion of LH by the pituitary rises. The rise in LH causes the ovarian follicle to disintegrate thereby releasing the ovum. The remnants of the follicle become a corpus luteum. The corpus luteum, under the influence of LH, produces another steroid hormone called progesterone. Progesterone further stimulates the thickening of the uterine wall. If the ovum is fertilised, progesterone helps the establishment and maturation of the foetus. If fertilisation does not take place the corpus luteum degenerates and progesterone secretion ceases. A detailed discussion of the regulation of the oestrous cycle is given in Chapter 2 of the core text.

Thyroid and parathyroid

thyroxine

calcitonin

The thyroid gland secretes thyroxine and related compounds. These hormones control the basic metabolic rate and also affect brain function. Lack of thyroid hormones leads to cretinism. The thyroid gland also secrete a hormone called calcitonin. This hormone depresses Ca^{2+} levels in plasma by reducing the mobilisation of bone calcium.

parathormone

The parathyroid glands are much smaller than the thyroid gland. They secrete parathormone which acts on plasma Ca^{2+} in the opposite direction to calcitonin.

Pancreas

glucagon
insulin

The pancreas secretes two hormones glucagon and insulin. A fall in plasma glucose concentration stimulates the secretion of glucagon. Glucagon stimulates the conversion of liver glycogen to glucose and also gluconeogenesis. This results in the release of glucose into the blood stream. Insulin has the opposite effect to glucagon. Thus insulin stimulates the removal of glucose from the blood stream into fat and muscle cells. In muscle, the glucose is converted into glycogen whereas in fat cells it is converted to triglycerides. Insulin secretion is stimulated by high plasma concentration of glucose and amino acids especially if these are derived from dietary intake. It is also stimulated by glucagon. Insulin and glucagon together not only control plasma levels of glucose but also effectively cause the transfer of glucose from the liver to muscle and fat cells.

α-cells
β-cells

It should be noted that the production of hormones by the pancreas is separated from the production of digestive enzymes produced by this organ. The endocrine function of the pancreas is confined to structures called islets of Langerhans. These contain two types of cells, the α-cells which secrete glucagon and the β-cells which secrete insulin.

Kidneys

renin

angiotensin II

The kidneys are mainly responsible for the regulation of the water and salt balance of the body fluids. A fall in Na^+ concentrations in the plasma coupled with a decline in blood pressure causes the kidneys to release an enzyme called renin. Renin causes the production of the hormone, angiotensin II, which in turn stimulates the formation of aldosterone by the adrenal cortex. Remember that aldosterone influences the reabsorption of Na^+ by the kidneys. Water balance, regulated by rate of urine production, is under the influence of ADH from the posterior lobe of the pituitary.

erythropoietin

The kidneys also secrete another important hormone, erythropoietin. This hormone promotes the production and maturation of red-blood cells. The production of erythropoietin is affected by the blood oxygen tension. Low oxygen tensions stimulate erythropoietin release which in turn stimulates red blood cell production thereby increasing the oxygen carrying capacity of the blood.

Stomach and duodenum

gastrin

secretin

cholecystokinin

The pyloric cells of the stomach are stimulated to secrete gastrin by the presence of food. This causes the stomach wall cells (fundic cells) to secrete hydrochloric acid. The acid juices leaving the stomach stimulate the duodenal cells to secrete secretin and cholecystokinin (also called pancrazymin). Secretin stimulates liver and pancreatic secretions especially water and ions. Cholecystokinin causes the gall bladder to contract and to force its contents out and into the duodenum. It also stimulates the pancreas to release its digestive enzymes.

Heart

atrial natriuretic factor, ANF

In 1981 cells in the walls of the atria were found to be in the source of a polypeptide hormone (circa 29 amino acids) subsequently called atrial natriuretic factor (ANF). Over 10 years ANF has been the subject of rigorous and intensive research. Certainly, atrial stretch as the upper chambers of the heart fill with blood is a major stimulus for ANF release into the bloodstream. At the kidney, the hormone promotes a natriuresis, greatly increased sodium and water excretion in the urine, through a variety of effects on glomerular and tubular function not yet fully understood. Thus ANF contributes to overall fluid and water balance along with ADH and aldosterone, and the heart becomes the latest addition to the list of endocrine organs.

9 Resume of major hormones

In Table 3 we have summarised the major hormones and their function. Many of these we have discussed in the text. You should realise, however, that there are many more hormones known than these listed here. Table 3 is very large, it might be helpful to try to learn it in sections. Start with, for example, the hormones produced by the adrenal glands, sex organs and thymus. By dividing this task up you will find it much simpler.

Hormone	Site of Production	Type	Control	Function
GHIF (or SS) (somatostatin) GHRF	Hypothalamus	peptide	GH levels	control of GH levels
TRF	Hypothalamus	peptide	T4 & T3 levels	TSH output
GnRH (GnRF)	Hypothalamus	peptide	gonadal function	LH/FSH output
CRH	Hypothalamus	peptide	cortisol/ACTH levels	ACTH output
PIF PRF	Hypothalamus	peptide	foetal development	prolactin output
Growth hormone (GH)	anterior pituitary	peptide	GHIF/GHRF	growth
Prolactin	anterior pituitary	peptide	PIF levels	milk production
TSH	anterior pituitary	peptide	T4 & T3 levels	T4 & T3 levels
ACTH	anterior pituitary	peptide	cortisol/TSH levels	cortisol output

Table 3 Cont'd over.

Hormone	Site of Production	Type	Control	Function
LH/FSH	anterior pituitary	peptide	relaxin/inhibin levels	ovulation follicle development & sperm development
ADH	hypoth./posterior pituitary	peptide	blood osmotic pressure	water reabsorption
Oxytocin	hypoth/posterior pituitary	peptide	suckling stretch of birth canal	milk ejection uterine contractions
Cortisol	adrenal cortex	steroid	ACTH levels	metabolism stress responses immune system
Aldosterone	adrenal cortex	steroid	renin/angiotensin	sodium reabsorption
ANF	cardiac atria	peptide	atrial stretch	sodium/water excretion
Adrenaline	adrenal medulla	tyrosine based	sympathetic nervous system	stress/metabolism
Testosterone	testis	steroid	GnRH/LH/FSH	male 2 sexual characteristics
Oestrogen/ Progesteron	ovary	steroid	GnRH/LH/FSH	female 2 sexual characteristics
Thyroid hormones	thyroid	tyrosine based	TRF/TSH	metabolism growth maturation
Calcitonin	thyroid para-follicular cells	tyrosine based	calcium levels in blood	calcium levels in blood
Parathyroid Hormone	parathyroid gland	peptide	calcium levels in blood	calcium levels in blood
Insulin	pancreas β cells	peptide	blood glucose levels	blood glucose levels
Glucagon	pancreas α cells	peptide	blood glucose levels	blood glucose levels
Gastrin Secretin Cholecystokinin	gastro-intestinal tract	peptide	digestive products pH	control of digestion motility
Thymosin	thymus	peptide	?	body rhythms ? sexual maturity
Melatonin	pineal	peptide	?	immune system
Erythropoietin Angiotensin II	kidney	peptide	blood oxygen tension Na^+	red blood cell maturation control salt balance

Table 3 Summary of the general properties of the major hormones. Abbreviations: GHIH - Growth hormone inhibiting hormone, GHRH - Growth hormone releasing hormone, TRF - Thyroid releasing factor, GnRF - Gonadotrophin releasin factor, CRH - Corticotrophin releasing hormone, PIF - Prolactin inhibitory factor, PRF - Prolactin releasing factor, ADH - Antidiuretic hormone, T4 - Thyroxine, T3 - Triiodothyronine, GH - Growth hormone, GIP - Gastric inhibitory peptide or glucose-dependent insulinotropic hormone, ANF - Atrial natiuretic factor, ACTH - Adrenocorticotrophic hormone, TSH - Thyroid stimulating hormone, FSH - Follicle stimulating hormone, LH - Luteinising hormone.

SAQ 10

Match the hormone with the appropriate function.

Hormone	Function
TRF	sodium reabsorption
oxytocin	red blood cell production
testosterone	hydrochloric acid release
erythropoietin	thyroid hormone secretion
gastrin	smooth muscle contraction
aldosterone	spermatogenesis

SAQ 11

Match the hormone with the site of production.

Hormone	Site of production
insulin	thyroid
calcitonin	pancreas α cells
erythropoietin	hypothalamus
cortisol	adrenal medulla
adrenaline	cardiac atria
TRF	anterior pituitary
luteinising hormone	corpus luteum
glucagon	pancreas β cells
progesterone	kidneys
ANF	adrenal cortex

Summary and objectives

This appendix has provided a summary of the organisation and function of the mammalian endocrine system.

Now that you have completed this appendix you should be able to:-

- describe the location of the main endocrine tissues;
- assign the major hormones to their site of production;
- assign functions to the major hormones;
- distinguish between the three classes (peptide, catecholamines and steriods) of hormones in terms of their secretion and uptake by target cells;
- identify the nature of secondary messengers from supplied data;
- explain, by using suitable examples, feedback control;
- explain how much hormone output is controlled by the hypothalamus and pituitary gland.

Appendix 3

Units of measurement

For historical reasons a number of different units of measurement have evolved. The literature reflects these different systems. In the 1960s many international scientific bodies recommended the standardisation of names and symbols and a universally accepted set of units. These units, SI units (Systeme Internationale de Unites) were based on the definition of: metre (m), kilogram (kg); second (s); ampare (A); mole (mol) and candela (cd). Although, in the intervening period, these units have been widely adopted, their adoption has not been universal. This is especially true in the biological sciences.

It is, therefore, necessary to know both the SI units and the older systems and to be able to interconvert between both sets.

The BIOTOL series of texts predominantly uses SI units. However, in areas of activity where their use is not common, other units have been used. Tables 1 and 2 below provides some alternative methods of expressing various physical quantities. Table 3 provides prefixes which are commonly used.

Mass (S1 unit: kg)	Length (S1 unit: m)	Volume (S1 unit: m^3)	Energy (S1 unit: J = kg m^2 s^{-2})
g = 10^{-3} kg	cm = 10^{-2} m	l = dm^3 = 10^{-3} m^3	cal = 4.184 J
mg = 10^{-3} g = 10^{-6} kg	Å = 10^{-10} m	dl = 100 ml = 100 cm^3	erg = 10^{-7} J
μg = 10^{-6} g = 10^{-9} kg	nm = 10^{-9} m = 10Å	ml = cm^3 = 10^{-6} m^3	eV = 1.602 x 10^{-19} J
	pm = 10^{-12} m = 10^{-2} Å	μl = 10^{-3} cm^3	

Table 1 Units for physical quantities

Concentration (SI units: mol m^{-3})

a) M = mol l^{-1} = mol dm^{-3} = 10^3 mol m^{-3}

b) $mg1^{-1}$ = µg cm^{-3} = ppm = 10^{-3} g dm^{-3}

c) µg g^{-1} = ppm = 10^{-6} g g^{-1}

d) ng cm^{-3} = 10^{-6} g dm^{-3}

e) ng dm^{-3} = pg cm^{-3}

f) pg g^{-1} = ppb = 10^{-12} g g^{-1}

g) mg% = 10^{-2} g dm^{-3}

h) µg% = 10^{-5} g dm^{-3}

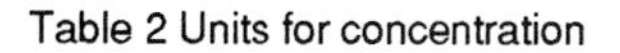
Table 2 Units for concentration

Fraction	Prefix	Symbol	Multiple	Prefix	Symbol
10^{-1}	deci	d	10	deka	da
10^{-2}	centi	c	10^2	hecto	h
10^{-3}	milli	m	10^3	kilo	k
10^{-6}	micro	µ	10^6	mega	M
10^{-9}	nano	n	10^9	giga	G
10^{-12}	pico	p	10^{12}	tera	T
10^{-15}	femto	f	10^{15}	peta	P
10^{-18}	atto	a	10^{18}	exa	E

Table 3 Prefixes for S1 units

Appendix 4

Chemical Nomenclature

Chemical nomenclature is quite a difficult issue especially in dealing with the complex chemicals of biological systems. To rigidly adhere to a strict systematic naming of compounds such as that of the International Union of Pure and Applied Chemistry (IUPAC) would lead to a cumbersome and overly complex text. BIOTOL has adopted a pragmatic approach by predominantly using the names or acronyms of chemicals most widely used in biologically-based activities. It is recognised however that there remains some potential for confusion amongst readers of different background. For example the simple structure CH_3COOH can be described as ethanoic acid or acetic acid depending on the environment or industry in which the compound is produced or used. To reduce such confusion, the BIOTOL series makes every effort to provide synonyms for compounds when they are first mentioned and to provide chemical structures where clarity and context demand.

Appendix 5

Abbreviations used for the common amino acids

Amino acid	Three-letter abbreviation	One-letter symbol
Alanine	Ala	A
Arginine	Arg	R
Asparagine	Asn	N
Aspartic acid	Asp	D
Asparagine or aspartic acid	Asx	B
Cysteine	Cys	C
Glutamine	Gln	Q
Glutamic acid	Glu	E
Glutamine or glutamic acid	Glx	Z
Glycine	Gly	G
Histidine	His	H
Isoleucine	Ile	I
Leucine	Leu	L
Lsyine	Lys	K
Methionine	Met	M
Phenylalanine	Phe	F
Proline	Pro	P
Serine	Ser	S
Threonine	Thr	T
Tryptophan	Trp	W
Tyrosine	Tyr	Y
Valine	Val	V

Index

A

B

C

L

M

N

O

P

Y

Z